Research Methodology and Data Analysis

Research Methodology and Data Analysis

The Future of Data Mining
Cem Ufuk Baytar, PhD
2022. ISBN: 979-8-88697-250-4 (Softcover)
2022. ISBN: 979-8-88697-315-0 (eBook)

Under the Paranormal Curve: Comparing Psychology Research Methods to Parapsychological Popular "Science"
Eric Charles Prichard
2022. ISBN: 979-8-88697-243-6 (Softcover)
2022. ISBN: 979-8-88697-343-3 (eBook)

Machine Learning Analysis of qPCR Data Using R
Luigi Marongiu, PhD
2022. ISBN: 979-8-88697-339-6 (Softcover)
2022. ISBN: 979-8-88697-380-8 (eBook)

E-Records Integrity Requirements
Orlando López
2022. ISBN: 978-1-68507-553-8 (Hardcover)
2022. ISBN: 978-1-68507-726-6 (eBook)

Predictive Analytics for Data Driven Decision Making – Tools and Techniques for Solving Real World Problems
B. Uma Maheswari, PhD, L. Ashok Kumar, PhD and R. Sujatha, PhD
2022. ISBN: 978-1-68507-650-4 (Hardcover)
2022. ISBN: 978-1-68507-770-9 (eBook)

More information about this series can be found at https://novapublishers.com/product-category/series/research-methodology-and-data-analysis/

Bruce J. West

West's Last Report

Is There an Optimal Funding Strategy for STEM Research?

https://doi.org/10.52305/LRKX7266

NOTICE TO THE READER

Library of Congress Cataloging-in-Publication Data

ISBN: 979-8-88697-608-3

Published by Nova Science Publishers, Inc. † New York

Contents

Preface

The Director of the Army Research Office[1] (ARO) sent me an e-mail on 10/29/2019 calling to my attention that senior Army leadership had the Single Investigator (SI) grants strategy of ARO under scrutiny. At the time I was an Army Senior Scientist having the title of ST in Mathematics with the protocol level of Brigadier General. There were about 40 such STs in various disciplines scattered throughout the Army. Part of an ST's duties is to provide scientific evaluations of technical problems identified by Army leadership. In this e-mail the ARO Director identified the following problem:

> The new Secretary of the Army is a proponent to reducing the number of grants we award to single institutions in favor of concentrating our efforts into more specific areas and awarding multiple grants to institutions. He suggested a while back to reduce the number of institutions by over half. No action has been taken to date.

He then posed a number of possible research questions to me along with the challenge:

---

[1]The Army Research Laboratory (ARL) is the Army's corporate research laboratory strategically placed within the Army Futures Command (AFC). ARO is a component of ARL whose primary responsibility is to fund and oversee extramural basic research that contributes to the Army's long-term operational capabilities. It drives cutting-edge and disruptive scientific discoveries that enable crucial future Army technologies and capabilities through high-risk, high-pay-off research opportunities. This is accomplished by funding investigative research into fundamental phenomena, which in turn provides the Army with key potential advantages in future conflicts.

> What is it that you can do over the next 3-6 months to answer these research questions and perhaps formulate a better model for segmentation of awarding SI grants to single versus multiple institutions?

I did not provide answers to his questions because at the time I believed that we did not have the measures with which to answer any well formulated list of questions, his, or anyone else's. In the following month I formulated a strategy to use the archived ARO data on grants awarded, papers published, honors received, etc. to construct the measures necessary to answer the questions on any such list. This strategy was put into the form of a 45 minute briefing and delivered to the ARO community to get broad–based feedback on the assumptions that went into the initial formulation. The intent was to bring the tools of science to the study of all things that could potentially enter into the formulations of measures of the quality of science research, including but not restricted to psychological, societal, historical, institutional and financial. The application of the Science of Science to Army supported research was proposed with the intended purpose of quantifying the quality of that research done by individuals members of research programs and within all manner of research institutions.

The Army is in the process of transforming how it supports, conducts, transitions and utilizes scientific research through the Army Futures Command (AFC). The intent is for science and technology to make the fighting forces more mobile, lethal and survivable, so they can successfully carry out today's less conventional missions and counter asymmetric warfare, as well as perform its more traditional functions. The financial support of science is intended to enhance the efficiency of transitioning a nascent scientific concept to the application workbench, then onto advanced technology development to design and build a prototype that can be worked into a fieldable system. These are the first five steps in the linear logical chain that was adopted by the military at the end of the second world war to transition basic research supported by the military to social and long-time military applications.

A white paper was received from RTI International in November 2019 to construct the necessary measures using the ARO data. Their experience with constructing such specialized measures and their working knowledge of ARO funding strategies convinced us that we could work together to reach our desired goal of quantifying the research quality using the individual SI, MURI and UARC strategies. The kick-off meeting was in August 2020 after the approval of the proposal, with a due date of deliverables at the end of April 2021. These deliverables included ways to improve the quality of research by providing institutional, programmatic and individual measures of research quality and determining the response of these measures to changes in funding protocols and scientific management, using data sets provided by ARO.

Acknowledgments

I thank the two remaining Army Research Office STs, Peter Reynolds and Stephen Lee, for their generous sharing of knowledge and experience in guiding me through my first and last effort of taking on the tasks of a PM, as well as for their suggestions on the interpretations of the research results of RTI International contained in Chapters 6 and 7. Also to my family, who nursed me through the triad of illnesses, a degenerative back condition, Parkinson's Disease, and 21 days of Covid. In particular Adranna and wife Sharon.

Abbreviations

AAAS: American Association for the Advancement of Science
AFC: Army Futures Command
AFOSR: Air Force Office Scientific Research
AIS: Author Influence Score
APS: American Physical Society
AR: Allometry Relation
AR: Applied Relevance
ARL: Army Research Laboratory
ARO: Army Research Office
BP: Breakthrough Potential
C-AR: Citation Allometry Relation
CDF: Cumulative Distribution Function
CI: Citation Index
CSIN: Collaborative Single Investigator Network
CLT: Central Limit Theorem
DARPA: Defense Advanced Research Projects Agency
DoD: Department of Defense
DEVCOM: Development Command
DSRSI: *Altmetrics* affiliate
DTIC: Defence Technical Information Center
EP: Empirical Paradox
FDE: Fractional Diffusion Equation
FoR: Field of Research
FPE: Fokker-Planck Equation
FRE: Fractional Rate Equation

GSE: Great Scientist Effect
IF: Impact Factor
IPL: Inverse Power Law
ISN: Institute for Soldier's Nanotechnology
IQ: Intelligence Quotient
MDF: Mobile Direct-Fire
MIT: Massachusetts Institute of Technology
MLF: Mittag-Leffler Function
MURI: Multidisciplinary University Research Initiative
NLM: National Library of Medicine
NRC: National Research Council
ONR: Office of Naval Research
OSRD: Office of Scientific Research & Development
PDF: Probability Density Function
PI: Principle Investigator
PM: Program Manager
RFP: Request for Proposal
SI: Single Investigator
SIC: Single Investigator Citation
SICN: Single Investigator Citation Network
SIP: Single Investigator Paper
SIPN: Single Investigator Paper Network
SSBN: SICN-SIPN Bipartite Network
SS: Scientific Significance
S2STEM: Science of STEM
STEM: Science, Technology, Engineering and Mathematics
SoS: Science of Science
UARC: University Affiliated Research Center
USMA: United States Military Academy
WoS: Web of Science
WWW: World Wide Web
YIE: Young Investigator Effect

Chapter 1

Research Quality

The American identity has nowhere been more carefully or positively analyzed than in 1835 by Alexis De Tocqueville in his remarkable book *Democracy in America* unless it is in the 2010 masterwork of Timothy Ferris *The Science of Liberty; Democracy, Reason, and The Laws of Nature* [1]. Both authors associate the roots of what it is to be an American with the underlying characteristics of Benjamin Franklin and Thomas Jefferson, who were both scientists and inventors. But Ferris traces the science upon which our modern view of liberty is based and transcends the arcane political notions of left and right, while setting the table upon which Vannevar Bush served up his report *Science, The Endless Frontier* [2], which provided a blueprint for government sponsorship of basic science.

Then as now, the arguments presented by Bush were not universally embraced and many of the fundamental concerns were anticipated in the 'Bush-Kilgore (BK) debates' [3]:

> Not everyone, however, was buying Bush's story. Starting in the early 1940s, Senator Harley Kilgore, a Democrat from West Virginia, championed a different national approach to science policy, one in which government investment would focus research and development directly on social goals and economic growth. A six-year political battle between Kilgore and Bush followed, to control not just US science policy itself, but, equally importantly, the rhetoric of science and progress. Bush, who had much of the lead-

> ership of academic and industrial science on his side, and who saw Kilgore as a threat to the independence of both elite academic science and the economic marketplace, became the decisive winner on both fronts: the 1950 bill creating the National Science Foundation gave scientists primary responsibility for determining the agency's research agenda.

The present work addresses a problem whose solution may guide the determination of the strategy for government supported research for a substantial fraction of the next century. The problem has to do with the concern expressed by Pentagon leaders regarding the manner in which the United States has been funding scientific research over the past few decades. In its simplest form the comments made by senior military leaders amount to the observation that government resources could be better used if we restricted support to only the top universities, since these institutions employ the most talented researchers, are home to the most promising students and house the best research facilities.

But is this observation true, or is it merely a pedestrian way of restating a commonly held but unsubstantiated belief of what ought to be true?

What is the scientific evidence that the top universities have the best of everything involving research? In addition, what is the scientific evidence that the top universities use selection criteria for their research questions that are more important to the military and the society which it serves than those made by supposedly non-top universities? The problem addressed in this essay is how to quantitatively critique assertions posing as answers to these and other similar questions, using measures that convince not just scientists of the truth or falsity of these assertions, but perhaps even more importantly, convince national leaders, as well as society as a whole.

One deceptively simple aspect of the problem is that investing limited resources in only the top schools sounds like such a reasonable thing to do. Given its evident reasonableness is probably the reason why no one has bothered to prove it. Even with the recent burst of activity into what has been called the *Science of Science*[1] there has been surprisingly little systematic effort to analyze

[1]Upon completion of this essay the excellent book *The Science of Science [4]* was published and although we treat similar material our purposes are very different. My concern, as stated, is a focused attempt to determine whether there is an optimum strategy for government funding

existing datasets and determine, once and for all, whether or not there is, in fact, an optimal strategy for the funding of scientific research.

For example, what would be the effect on the quality of U.S. science of reducing the number of institutions the government supports through single investigator (SI) grants by one-half, or even more? I do not know of anyone who can show with a quantitative model, that is to say with reliable measures, the answer to that question. We may have suspicions based on our experience, intuition if you will, but scientists have a higher standard than that set by a reasonable argument, or a historical precedent, such as used in the historical debates alluded to above. More importantly, how are we to influence the mindset of senior national leaders, without appearing to be self-serving partisans for science, wanting to do science solely for its own sake. The only way available to scientists is to shine the unforgiving spotlight of scientific analysis and reason on available datasets and lay out in plain view what the analyses of these datasets tell us. We must apply the same techniques that have proven their value in reaching innovative solutions to the nation's technical problems to determining the quality of the funding strategies to that same scientific research.

1.1. The Problem

A news article appeared in the global magazine *Foreign Policy* in October of 2018, written by the pentagon correspondent L. Seligman, in which the Pentagon was headlined as being criticized for their 'spray and pray' approach to scientific innovation. 'Spray' is intended to connote the less than optimal distributing of resources across the United States for scientific investigations and 'pray' is meant to characterize the method by which the results of those funding investments are evaluated. The cleverly worded pejorative phrase *spray and pray* was apparently invented by a high level tech entrepreneur to depict the large number of small investments in commercial technical projects made by the Pentagon. This could be dismissed as self-serving sarcasm if it did not have at least the tacit endorsement of some senior leaders within the Pentagon. The implied view is that to win in our competition with China and other adversary nations to be the world's scientific leader we need to identify the best universi-

of scientific research. The more expanded scope of *[4]* was to use the quantitative methods of science to answer more fundamental questions of science.

ties in the U.S. and give each of them a substantial grant that is sufficiently large to attract the kind of venture capital funding that can sustain a small start-up." But is this the best way to address the scientific questions of importance to a country or its military?

My response is that I do not have a definitive answer to such questions. But then neither does the cocksure entrepreneur being quoted. There are no lack of opinions. They run the gamut from one extreme: "If you want the best scientific innovation you must invest in the individual scientist." to the other: "The best innovation consistently comes from the best institutions and that is where you should put your money."

On the other hand, I do know that datasets do exist from which, using properly defined data processing techniques, these and other similar questions may be answered in whole or in part.

The answers to the above question and all similar questions boil down to what evidence can be determined in the form of quantitative measures and their qualitative interpretation, based on the analysis of available datasets to support or refute a well articulated form of the stated concern. Presumably, most if not all of the data necessary to provide at least preliminary answers to these questions are archived in the computer facilities of ARO and in other government, as well as in non-government, computer archives. The scientific white papers, proposals and those projects receiving grants, the published papers supported under these grants, the number of citations each such paper received over the time of the grant, etc., are all there. However, we need to define the measures that enable the transformation of these datasets into information. It is the patterns in the resulting information and the subsequent interpretation of those patterns that can lead to knowledge about the efficacy of alternative funding strategies.

Of course, what a scientist finds convincing would not necessarily win over a national leader, due in no small part to what they each think is important, and how they prioritize that importance. It is in the identification and interpretation of programmatic patterns that we can use to determine the value of specific funding research strategies to the Army. We emphasize here that we are restricting our view to that of science, technology, engineering and mathematics (STEM) research supported by the Army because the ARO is the primary data source used in the analysis in a recent project we subsequently discuss. The question is: Can we justify the existing ARO funding strategy of identifying and supporting the SIs with the best STEM ideas regardless of their institutional affiliations? Moreover is this an optimal strategy for identifying and carrying out STEM research of most relevance to the Army?

The *Foreign Policy* article discussed above reported the position of a number of leading industrial research companies that funding individual researchers located at geographically disperse locations is not an efficient way to support STEM research. These venture capitalists propose what they believe to be a much more rational method for distributing resources to support STEM research, that being to take the government's budget for STEM research and divide it into a small number of pieces, each of which could support a relatively large research activity. Any one of these large activities could attract 'startup funding' from venture capitalists. The reporter went on to juxtapose quotes from General Murray concerning the AFC with those of entrepreneurs, giving the impression that senior Army leadership is sympathetic to such a Draconian change in research support policy. For example:

> (GEN) Murray touted the establishment of Army Futures Command, a new modernization effort that boasts lofty goals of leveraging innovation in the private sector, and said the group plans to partner with a number of tech companies on experimental pilot programs.
>
> "In many ways, Army Futures Command is a start-up in itself, so we are trying to cast a very wide net," Murray said during a press conference at the event.
>
> But Stephens said this strategy (of)... "Moving away from the 'spray and pray' model to a concentrated approach will help build U.S. businesses that can compete and win against the traditional major defense contractors..."

Of course, one could address these arguments point by point, but that would not serve our present purpose because it would be rhetoric and debate, not science. Debate and discussion are not to be avoided when the end goal is to resolve confusion and bring clarity to a situation, but all too often what is achieved is the formulation of a well articulated argument in support of bias and preconceptions with little or no concrete evidence. The purpose here is to determine what science can tell us about which is the more viable way to distribute resources in support of STEM research. For generality in achieving this undertaking we focus on the construction of quantitative measures of the quality of research, pertaining to research organizations, such as universities or government laboratories; research programs, that are directed and funded by government agencies; as well as, the careers of individual researchers. Only in this way can the questions surrounding the proper support of STEM research be resolved, particularly in the quantification of the quality of that research supported by the U.S. Army.

Consequently, to answer these last two questions, as well as other related questions, we postulate three scenarios against which to test the measures we subsequently develop and use to quantify the quality of Army-supported STEM research.

Funding Scenarios:

The first scenario addresses the distributed funding strategy in which requests for proposals (RFP) are competed for by researchers at all universities across the U.S.. The RFPs solicit responses from single investigators (SIs) through the broad agency announcement (BAA) which contains the thematic areas of research interest to the three services. The ARO, the Air Force Office of Scientific Research (AFOSR) and the Office of Naval Research (ONR) each has its own particular scientific areas that are emphasized in accordance with its mission domain. The data compiled, including the number of publications, citations, honorary awards, etc., by each of the services over the years is available to be used to directly address the *spray and pray* criticism mentioned above. Given that the author was the Senior Scientist in Mathematics at ARO from June of 1999 until retirement on 31 July of 2021 it was natural to use the data at hand to address the Pentagon's concerns. This will be discussed in some detail subsequently.

Each of the ARO program managers (PMs) identifies and invests in the most innovative research ideas proposed by SIs and which accounts for the bulk of ARO's Core Research Program. Grants are awarded by the PMs through a competitive selection process and are each led by a Principal Investigator (PI). A typical grant supports a project designed to be completed within three years with a budget of approximately $120K per year on average.

The second scenario, in contrast to the first, focuses on a localized funding strategy in which a relatively large research grant is awarded to a top university to form what is called a University Affiliated Research Center (UARC). A UARC is a DoD strategic research center associated with an university, which supports multiple STEM departments whose researchers, both singly and collectively, contribute to the strategic thematic area of the UARC. Here too the PM has the final word on the acceptance or rejection of a new project supported by the UARC, but the pool of proposed projects is restricted to members of the UARC, thereby limiting the influence of the PM on the program. Of particular interest to us here is the *Institute for Soldier's Nanotechnology* (ISN) at MIT which provided data for the study which we subsequently use to address the questions posed.

The UARCs are university–led collaborations among universities, industry and Army laboratories that conduct basic, applied and technology demonstration research. These centers are part of a large strategic funding initiative at DoD and each one is funded by a cooperative agreement awarded to a particular university that establishes a long–term, large–scale research center developing fundamental capabilities in a key domain of basic research. Each UARC receives a five-year award, which is renewed based on in-depth performance reviews, with a budget of approximately $5M per year on average.

The third funding scenario combines aspects of both the distributed and localized funding strategies. It is a hybrid having aspects of both the distributed SI strategy and the localized UARC strategy, and is the Multidisciplinary University Research Initiative (MURI). A MURI project has a unique and lengthy procedure for developing its RFP, which typically identifies a research problem that is at the nexus of two or more disciplines and has published indications of its being ripe for a more concentrated influx of resources. The MURI supports research by teams of investigators whose research is a collaboration among the traditional STEM disciplines and most MURIs involve researchers from multiple STEM departments and often includes collaboration among scientists at multiple research institutions. As in the other funding strategies a MURI proposal undergoes a rigorous evaluation process of academic review with the PM making the final recommendation for funding. A MURI is a tri–service Department of Defense program having a broad multidisciplinary thematic focus and is awarded approximately $1.25M per year for no more than five years.

Datasets from each of these three funding scenarios are herein used to develop measures of research quality with which we can compare and contrast the different funding strategies. So what are the measures we propose to use? Why these measures and not others? The answers to these and related questions constitute the bulk of the discussion presented in the following chapters. However, before we address these issues we lay the ground work for what constitute answers from a historical perspective.

1.2. Forecasting Technology

We humans typically have an aversion to uncertainty. We complain about being in a rut, or how boring the routine of our lives have become, and fantasize about getting away from it all, but the truth is that most of us are comfortable with being able to predict our futures. Change makes people nervous, whether this is the result of facing an unknown future, a consequence of not being in control, or a combination of these and other factors is unclear. Unforeseen change, in particular, disrupts what has been laid out with no promise of comfort or success. Consequently, parents plan the future for their children and cry and moan when their beloved offspring deviates from the plan and transfers from an engineering to a fine arts major, or the other way around. All of this is by way of asserting that humans not only prefer to have a planned future, but at a minimum they want to have the broad outline of their future visible and under their control. Accepting this as the preferred mode of living for most individuals it is not unexpected that we place a premium on prediction and forecasting, both individually and collectively. For the latter, we turn to leaders that for one reason or another we believe we can trust.

Business, political, and military leaders are interested in forecasting when the next technology breakthrough will occur, in what domain of activity it will occur, how much it will cost, how rapidly it can be fielded, and in the answers to seemingly countless other similar questions. The motivations leading to these question in each of these categories are almost never the same in detail and the science on which the technology is based is nearly always irrelevant.

Business leaders need to forecast when the next and newest widget will emerge in order to maximize the profit resulting from being the first to market, as well as continue in that profiting before the competition innovates widget-2. This insures the rise in the company stock, their bonus and continuing to climb the ladder of success.

Political leaders require the acumen to foresee which legislative action is needed to facilitate or inhibit the growth of new technologies in the private sector, whichever will result in their re-election and maintain the continuity of political power.

Military leaders anticipate the technological breakthroughs needed to ensure that the force under their command stays ahead of their country's adversaries, in areas such as weaponry, training and intelligence. Specifically, how to assign limited resources to realize the desired outcomes.

The lack of a theory of technology forecasting is a major difficulty when it comes time for these various leaders to answer questions surrounding how to make technological forecasts. Without such a theory to provide the intellectual tools and language with which to think about the development of ideas within a discipline, it is prohibitively difficult to solve problems within that discipline, much less to make forecasts concerning breakthrough applications. Without such a theory it is nearly impossible for this diverse set of leaders to communicate with one another concerning potential breakthroughs within a particular discipline, and certainly not across disciplines, in any meaningful way.

Let's take a moment and consider what all this means using an example. A well-educated person of the early nineteenth century may have been able to imaginatively extrapolate an existing technology to a new application, say applying a steam engine to developing new modes of transport over land or water. However without an empirical language to relate the storage of steam power in a different readily usable form, for example, in batteries, that same person would find the electrical lighting of a city to be a mystery, close to magic. The same can be said generally of technology language limitations within a society that restrict the growth and development of technology.

Three centuries ago Sir Isaac Newton (1642-1726) transformed *Natural Philosophy* into what we mean today by *Science* through the introduction of the quantitative mathematics of motion into physics and thereby changing the pursuit of wisdom into the pursuit of knowledge. Newton's life spanned the transition from the 17th to the 18th century, when wars were fought with long bows, swords, cavalry, muskets and cannon. These images are intended to emphasize the contrast between the tools used by today's scientific leaders and modern weapons of war with those of Western civilization of the 18th century. But more than that, it is to draw attention to the difference in the ways we think about science and the technology of warfare. It is not just that muskets have transitioned into AK-47s, horses into UAVs and horse-drawn cannon into Abrahms tanks, but it also concerns how society developed and the ways in which 'social systems' were developed to support the science that makes these new weapon systems not just possible, but also socially acceptable.

One way to stress the differences in how we think today relative to three centuries past is by examining how the technology has evolved over time and clarify what we mean by the forecasting of technology innovation. A focused version of technology forecasting is typified by the time it takes for the 'quality' of a given developing technology to double. The doubling time of the thermal efficiency of hydrocarbon–fueled stationary steam power plants from the middle 1700s to the middle 1800s was 33 years [5]. The measure of quality in this case is taken to be efficiency, as measured by the ratio of the generating cost to the price of the subsequent benefit derived. Another measure of quality is Moore's Law, which predicts a doubling of the density of transistors packed onto a microchip every 18 months. This latter functional relation has held up since the 1960s and enabled companies, as well as governments, to reliably invest time and resources in the digital economy on the short term with a cost-benefit analysis that has rarely been found in predicting the capability of developing technologies in previous centuries. The contracted time scales for the development of this technology is a reflection of the oil lamps, coaches and mail delivery systems of the 18th and 19th centuries compared with the electric grids, jets and internet of the modern world.

The kind of exponential growth in the quality (more for less) of a technology can be identified in the historical record of technology development and led John Lienhard to make a number of surprising empirical observations. Professor Lienhard [6] argued that the quality of a given technology follows a path of exponential growth over time at a constant rate, or it did prior to the middle 1830's. He noted that at the inception of each new technology the growth rate appears to be fixed and to remain constant throughout the growth phase of the technology. Cardwell [7] acknowledged the rapid convergence of science and technology in the 19th century and quoting Whitehead noted:

> The greatest invention of the 19th century was the invention of the method of invention....

This observation by Cardwell suggests a second level of technology evolution. This second level of evolution, in fact, involves the interaction of science and technology, which apparently did not formally exist prior to the mid-1830s [8]. The first level is what we have discussed until now and is concerned with incrementally improving the technology in steps by which a stated technology purpose is achieved based on a given scientific principle. However, a given technology at some point in time becomes exhausted, or complete. To go beyond that point of completion in achieving the same technological objective it becomes necessary to shift scientific principles. An example of this is in the domain of transport of goods in support of a growing society. One can readily trace through time the tonnage hauled by beasts of burden (living haulers), later by canals interconnecting inland water ways (boats with steam engines), only to be replaced by railroads (steam followed by diesel engines), which in turn yielded substantial market share to 18-wheeler trucks on the interstate highway system (internal combustion engine). Each transport technology was replaced by one dependent on a new scientific principle, or an old technology newly adapted to a new transport domain.

The second level of technology growth is more encompassing in its effect in that it addresses how the process of technological evolution may in fact change over time. Lienhard expanded on Cardwell's arguments and summarized his insights into a single sentence which we refer to as the *Lienhard Hypothesis* ($\mathcal{LH}$) [5]:

> It therefore appears that what really happened in the early 19th century was that, as attention was shifted from actual inventions to the method of invention, that the quality of the method was the thing that began to improve exponentially.

A recent rediscovery of the $\mathcal{LH}$ was made by Kott [9] using data from a collection of diverse weapons systems called mobile direct-fire (MDF) systems. Kott discovered a parsimonious regularity describing the growth in these vastly different technologies over time. The multiple and widely different technologies – from a bowman to a tank – when expressed in the proper representation, gives rise to a regularity measure that clusters around a simple curve. The theoretical basis for this empirical curve was developed shortly thereafter by West et al. (WWK) [10] who applied the mathematics of scaling to nearly a millennium of

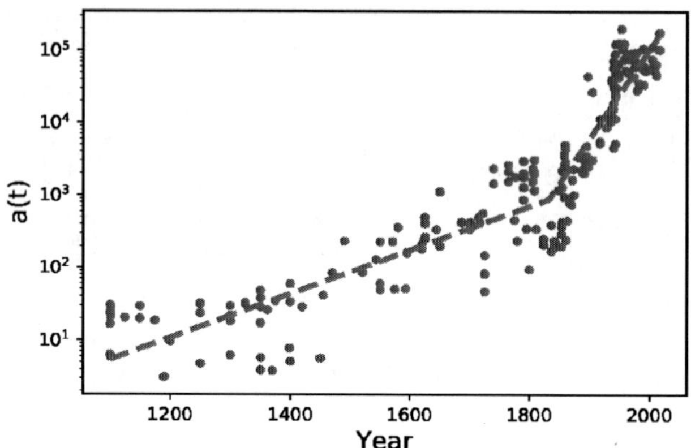

Figure 1.1. The graph depicts the growth of the average kinetic energy per unit time delivered by a MDF to the target over nealy a millenium. The kink in the curve at the year 1832 supports the $\mathcal{LH}$, as discussd in the text. From [9] with permisssion.

MDF data, stretching from 1100 to 2015 CE as depicted in Figure 1.1.

Note that the dramatic change in slope of the MDF technology growth curve at the year 1832 obtained in WWK is precisely the change in growth rate anticipated by the $\mathcal{LH}$. It had also been previously observed by Kott [9], who had used what he referred as the maximum kinetic energy of projectiles that an MDF system can direct at a target/unit time/unit mass of the total system as the 'primary factor' to produce a figure equivalent to Figure 1.1. Kott and subsequently WWK used a fitting curve to determine the minimum number of important variables with which to model the data, including the discontinuity occurring in the rate of technology innovation in the MDF systems. This is the first theoretical support of the $\mathcal{LH}$ that incorporates societal influence into the evolution of technology [10].

Almost every technology prior to the decade of the 1830s had a doubling rate of approximately 30 years. During that decade the doubling rate abruptly increased, thereby shortening the time interval between successive incremental developments in a new technology. This happened, according to Lienhard [6], because during that decade Western society institutionalized technology and its evolution and it was only in the immediately preceding decades that society recognized engineering and science as professions.[2]

1.3. Truncated History of Military Research

Prior to World War II there was no large-scale scientific effort supported by the federal government. In today's terms a large-scale effort would be one that is a measurable percent of the federal budget. Towards the end of World War II, Vannevar Bush, who had presided over the government research effort since the beginning of the war as Director of the Office of Scientific Research and Development (OSRD) (1941-47), responded to a request from President Roosevelt regarding how to transition war research to the private sector with the report *Science-The Endless Frontier* [2]. In this now legendary report he argued that the United States, in order to retain the STEM advantage achieved during the war years, needed to build a civilian-controlled organization for fundamental research to be conducted at universities with close liaison with the Army and Navy to support national needs and with the ability to initiate, conduct and carry basic research to completion.

He emphasized that STEM research is most successful in realizing breakthroughs when done in an atmosphere free of the adverse pressures of prejudice, convention, or short-term commercial interests. This absence of hierarchical structure stands in sharp contrast to military tradition in which he argued it should be embedded. V. Bush believed it possible to retain an outside alternate organizational structure, but working in close collaboration with the traditional military structure. Such a civilian controlled organization could foster and nurture science and its application to new technologies, through engineering. In Bush's own words [2]:

> ...such an agency ... should be ..... devoted to the support of scientific research... Industry learned many years ago that basic re-

[2]On March 16, 1802, President Thomas Jefferson signed the *Military Peace Establishment Act*, directing that a corps of engineers be established and "stationed at West Point in the state of New York, and shall constitute a Military Academy." Thus, the USMA came into existence as the first engineering school in the newly formed United States of America. On the other had, in 1861 MIT was established as the first school in the US with a full engineering ciriculum.

> search cannot often be fruitfully conducted as an adjunct to or a subdivision of an operating agency or department. Operating agencies have immediate operating goals and are under constant pressure to produce in a tangible way, for that is the test of their value. None of these conditions is favorable to basic research. Research is the exploration of the unknown and is necessarily speculative. It is inhibited by conventional approaches, traditions and standards. It cannot be satisfactorily conducted in an atmosphere where it is gauged and tested by operating or production standards. Basic scientific research should not, therefore, be placed under an operating agency whose paramount concern is anything other than research.

His vision materialized through the development of the ONR in 1946, the ARO in 1951 (as the Office of Ordnance Research), the AFOSR in 1950 (as the Air Research and Development Command), and the National Science Foundation (NSF) in 1950; albeit, none of these organizations followed all his suggestions regarding the management of scientific personnel and the support of science. These agencies continue to support research conducted on university campuses and while these agencies were being established a substantial number of research laboratories were stood up by the military services to house government scientists and engineers that would collaborate with academic and industry STEM investigators. His cautionary words concerning the incompatibility of basic STEM research and mission agencies turned out to be prophetic.

However, the dire consequence of the incompatibility of mission and research was held at arm's length for over 50 years by a set of checks and balances put into place in order to insulate basic research (6.1) from the pressures of applied research (6.2). The separation of the research being supported through the military services, but done on university campuses (basic) from the research being done within government laboratories (both basic and applied) has maintained, by and large, the separation between fundamental and applied research, see for example, West et al. [11] for a fresh perspective on this on-going debate. However, V. Bush's cautionary words have been echoed [12] by members of the National Research Council (NRC). Congress directed the DoD to have the NRC study the nature of the basic research being funded by the DoD. The findings of the report of most relevance to the present discussion are [12]:

> A recent trend in basic research emphasis within the Department of Defense has led to a reduced effort in unfettered exploration, which historically has been a critical enabler of the most important breakthroughs in military capabilities.
>
> Generated by important near-term Department of Defense needs and by limitations in available resources, there is significant pressure to focus DoD basic research more narrowly in support of more specific needs.
>
> The key to effective management of basic research lies in having experienced and empowered program managers. Current assignment policies and priorities (such as leaving a substantial number of program managers positions unfilled) are not always consistent with this need, which might result in negative consequences for the effectiveness of basic research management in the long term.

Two studies [13, 14] focused on the efficacy of the 700 laboratories and research centers constituting the Federated Laboratory System. John H. Hopps Jr., the then Deputy Director of DoD Research & Engineering and Deputy Undersecretary of Defense, observed that our "defense laboratories should have the same attributes as our transformed uniformed military forces." He specifically pointed out that STEM research should share the characteristic of the modularity of the joint forces with the parallel attributes of [14]:

> ....productivity; responsiveness and adaptability; relevance, programming, and execution and application; and perpetuation of knowledge.

In opposition to this notion of modularity it is interesting to recall V. Bush's remarks [2]:

> Science is fundamentally a unitary thing....Much medical progress, for example, will come from fundamental advances in chemistry. Separation of the sciences in tight compartments....would retard and not advance scientific knowledge as a whole.

The variability in research and creativity is fundamentally at odds with uniformity and regimentation. One does not create on demand; it is the exploration of new alternatives, going down blind alleys, and even failing that is at the heart of innovative research. Again quoting from V. Bush [2]:

> Basic research is a long-term process - it ceases to be basic if immediate results are expected on short-term support.

The impact of these verbal arguments has diminished over the years, resulting in a number of positions regarding the utility of how government ought to support research, even the very need for basic research by the military services has been called into question. The anecdotal evidence in support strategy of V. Bush is countered with other equally impressive anecdotes that seem to contradict it. Thus, we arrive at a stalemate at the level of rational discourse using historical precedents and must find resolution to this conflict elsewhere.

The United States emerged from World War II as the world's leader in research, after which it initiated the experiment in wide-spread, government-funded, civilian-controlled scientific research that is now in question. After more than 70 years, with the changes in social pressures and goals, the military services have taken on multiple roles in modern warfare. It is past time to reexamine the management and distribution of the resources necessary to maintain the world's leadership role in STEM research.

1.4. Is Everything Normal?

It is important to have a clear idea of the role statistics play in how we understand the quality of STEM research. Statistics was introduced into science to clarify the disruptive aspects of nonsimplicity in constructing predictions of how a process behaves in making its way from its present state to a future state. We therefore spend some time in critiquing the use and misuse of statistics, not in discussing its mathematics. There have been a variety of programs touted for the determination of quantifying the success of preforming a given task. One such program is Six Sigma. The name Six Sigma is taken from the

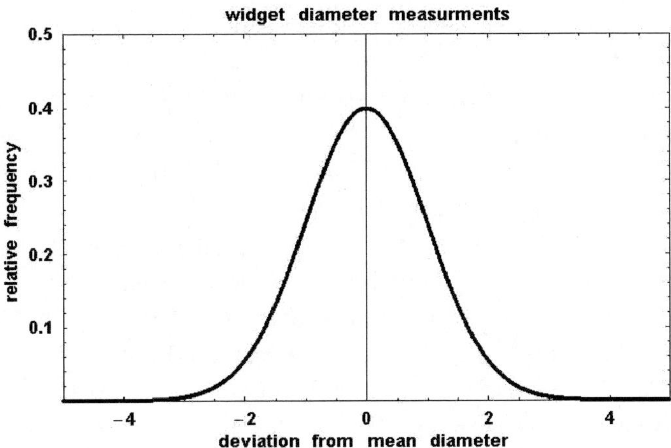

Figure 1.2. The normal (Gauss) distribution is depicted in units of the standard deviation σ centered on the average value of the random variable. Any random variable with normal statistics, expressed in such normalized form collapses onto this curve.

bell-shaped probability density function (PDF) shown in Figure 1.2, and first[3] constructed by the polymath Johann Carl Friedrich Gauss (1777-1855), to quantify the variability observed in the results of experimental measurements taken in the physical sciences. This PDF was necessary to explain the historic mystery that two ostensibly identical experiments never yielding the same results, and yet it is the reproducibility of experiments that makes science viable and therefore valuable. It is determined that there is always a measurable amount of random variability in the physical observable, which is not anticipated by the mechanics of classical physics.

In the world view of Gauss the average value of an observable is the single most important quantity for characterizing a phenomenon and the fluctuations around that average value are due to errors of measurement and/or random fluctuations of the environment not taken into account in a prediction. The parameter sigma (σ) quantifies the typical magnitude of these fluctuations and in

[3]The same year, 1808, an American mathematician Adrian [15] obtained the same mathematical resolution to the mystery.

statistics it is called the standard deviation. Sigma determines the width of the distribution, so the smaller the value of σ relative to the average, the less the variability and the better the average represents the data. Consequently, in this view, the world consists of networks made up of linear additive processes, whose random variability is described by a bell-shaped probability density curve, centered on an average value, with a width determined by sigma.

The Gauss model is very useful in manufacturing, where the specifications for the production of a widget can be given with as high a tolerance as can be achieved on a given piece of machinery used in its production. Of course, no two widgets coming off the production line are exactly the same, there is always some variation. Suppose that the widget can be characterized by a diameter and the variation around the average value of the measured diameter is given by a bell-shaped curve. If all the measured diameters are divided by the standard deviation in a given production run then the new variable is expressed in units of the standard deviation, σ.

In Figure 1.2 the results of producing this hypothetical widget are shown. One of the major attractions of this approach is the universality of the shape of the distribution where the standard deviation is the unit of measure. All processes that have Gaussian, or Normal statistics as they are called by mathematicians, can be superimposed on the given curve. A well-known property of the Normal (Gauss) distribution is that 68% of all widgets produced fall between plus and minus one sigma; 95% of all widgets produced fall between plus and minus two sigma; 99.7% of all widgets produced fall between plus and minus three sigma and so on.

The Six Sigma program maintains that the variability seen in Figure 1.2 is an undesirable property in the production of widgets and should be reduced to zero, which means achieving the condition: *average value* $>> \sigma$. Uniformity of outcome within the specified tolerance level is the goal of management in the process of manufacturing. To make what is implied by the PDF transparent assume that one million widgets are sold and 99.7% of them are within the specified tolerance limits. On its face one might be convinced that this is very good, with the production line functioning at the three-sigma level. But on closer examination it also implies that the company will have 3,000 unhappy customers. If the widgets were vaccines slated to be used by the general population during an epidemic, a three-sigma level would not be acceptable. A six-sigma level of 99.9997%, with its approximately 30 faulty doses of the vaccine out of the million shipped would be a more acceptable number.

So this is what the Six Sigma program is all about. How to take a manufacturing plant, an organization, or any other large-scale social activity whose output has an unacceptable level of variability and reduce that variability to a six-sigma level; variability is assumed to be bad and uniformity is assumed to be good. This is obviously a desirable goal in the manufacture of widgets, but it could mean disaster if this thinking is applied in an inappropriate context, specifically one in which variability is a desirable property.

When first introduced the six-sigma argument was attractive, with simple, but acceptable, assumptions about the process being measured and ultimately controlled through the management of variability. After all, we have been exposed to arguments of this kind since we first entered college and experienced that every large class was graded on a curve. That curve invariably involves the Normal distribution, where as shown in Figure 1.2 between +1 and -1 lie the grades of most of the students. This is the C range, between plus and minus one σ of the class average, which includes 68% percent of the students. The next range is the equally wide B and D interval from one σ to two σ in the positive and negative directions, respectively. These two intervals capture another 27% percent of the student body. Finally, the top and bottom of the class split the remaining 5% equally between A and F (or is it E?).

In the social sciences the bell-shaped curve introduced the notion of a statistical measure of performance relative to some goal. An organization, or network, is said to have a quantifiable goal, when a particular measurable outcome serves the purpose of the organization. A sequence of realizations of this outcome produces a dataset; say the number of research articles published per month by members of a research laboratory. Suppose a laboratory's ideal publication rate (goal) is specified in some way. The actual publication rate is not a fixed quantity, but varies a great deal from month to month and year to year. If 68% of the time the lab essentially achieves its goal, the lab is functioning at the one-sigma level. If another lab has 95% of its realizations at essentially the ideal rate, it is functioning at the two-sigma level. A third lab, one that seems to be doing extremely well, delivers 99.7% of the ideal rate and is functioning at the three-sigma level. Finally a six-sigma lab delivers 99.9997% of its publications at, or very nearly the ideal rate [16].

What six-sigma means is that the variability in the publication rate has been reduced to nearly zero and based on a linear additive model of the world, this ought to be both desirable and achievable. The logic of assessing the quality of the laboratory research is the same as that used to assess the quality of the students, or of an assembly line, and relies just as much on the bell-shaped curve. However, when we examine empirical data this interpretation of the world is not only called into question, but is found to entail decisions leading to catastrophic consequences.

Typical is not Normal

We now examine a situation in which the Normal PDF is universally accepted without question and yet is demonstrably wrong. Gupta and co-workers analyzed the achievement tests of over 65,000 students graduating high school and taking the university entrance examination of *Universidade Estadual Paulista* (UNESP) in the state of São Paulo, Brazil [17]. In Figure 1.3 some of the results of the entrance exam are recorded for high- and low-income students. It is clear that the humanities data in Figure 1.3a seem to support the conjecture that the Normal PDF is appropriate for describing the distribution of grades in a large population of students. The solid curves are the best fits of a Normal PDF to the dataset, and gives a different mean and width for public and private schools. In the original publication the data was represented in a variety of ways, such as between the rich and poor, but that does not concern us here, since the qualitative results turned out to be independent of the way in which the dataset is partitioned.

In Figure 1.3b the data from the physical sciences is graphed under the same grouping as that of the humanities in Figure 1.3a. One thing that is clear from a visual inspection of both partitionings of the dataset is that the PDF of student achievement in the physical sciences is remarkably different from the bell-shaped curve. The distribution of grades under the same groupings for the biological sciences depicted in Figure 1.3c is qualitatively indistinguishable from that of the physical sciences, which is to say it is not fit by a bell-shaped curve, but also has a long tail.

The first thing to notice is that the distributions of grades for the STEM disciplines in general are nothing like those in the humanities. The distributions

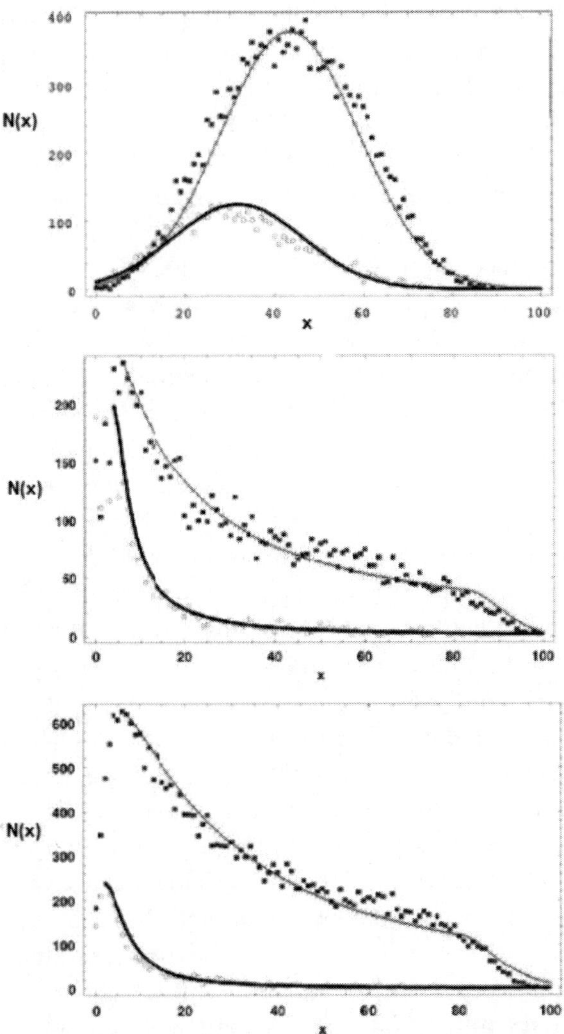

Figure 1.3. The distribution of grades for 65,000 students taking the university entrance examinations to the Universidade Estadual Paulista (UNESP) in the state of Sao Paulo, Brazil. The number of students $P(n)$ receiving a grade n is depicted for day students (upper curve) and night students (lower curve) in each panel: (a) humanities, (b) physical sciences and (c) biological sciences. The solid lines give the best IPL fit to each of the data sets. From [16] with permission.

are so different that if they were not labeled it would be impossible to determine by inspection that they refer to the same general phenomenon of learning. So why does normalcy appear to apply to the humanities and not to the sciences?

One possible explanation for this difference in the grade distributions between the humanities and the sciences has to do with the structural difference between the two learning categories. The humanities consists of a collection of disjoint groups of disciplines including language, philosophy, sociology, civics and a number of other relatively independent areas of study. We use the term independent because what is learned in sociology is not necessarily dependent on what is learned in philosophy, but is at most weakly dependent on what is learned in language studies. Consequently, the grades obtained in each of these separate disciplines are essentially independent of one another, thereby satisfying the conditions of Gauss' two hundred year old argument. In meeting Gauss' conditions the PDF of grades in the humanities takes on the bell-shaped form of normalcy.

On the other hand, every STEM discipline builds on previous knowledge. Elementary physics cannot be understood without arithmetic, algebra and geometry and the same could be said for biology. More advanced physics cannot be understood without the differential calculus, vectors and even elementary physics. Similarly, understanding biology requires some mastery of chemistry and physics. The different STEM disciplines form an interconnecting and ascending web, starting from the most basic and building upward, a situation that violates Gauss' assumptions of independence and undercuts the idea that the average value provides the best description of the process. The empirical distribution of grades in STEM disciplines clearly show extensions out into the tail of the PDF with no clear peak and consequently no characteristic scale with which to characterize the dataset. Consequently, the average values, so important in Normal statistical processes, become almost irrelevant in nonsimple phenomena, such as the learning of science. A better measure than the average with which to characterize a nonsimple process is one that quantifies how rapidly the tail of the distribution falls towards zero.

The distinction between the distribution of grades in the humanities and STEM disciplines is clear evidence that the Normal PDF does not describe the typical situation, at least not in STEM. The bell-shaped curve of grades is imposed through education orthodoxy and accepted by our preconceptions. The evidence clearly shows that the Normal PDF is not indicative of the process by which students master STEM information and knowledge. Thus, the pursuit and achievement of intellectual goals, whether in science or engineering, is not normally distributed. A recent book [18] details how this blind acceptance of averages as a way to characterize students has hobbled the U.S. educational system.

Consequently, the Six Sigma program, being crucially dependent on Normal statistics, is not applicable to nonsimple phenomena such as learning in general, nor to STEM research in particular. The variability targeted for elimination by the Six Sigma program under the assumption that the variability is described by Normal statistics is invalidated by the long-tailed PDF observed in truly nonsimple processes. The tails indicate an intrinsic variability that is part of the internal dynamics of the process and is not contained in the simpler processes where Normal statistics are valid. Moreover working to reduce the tail region may in fact remove the very property that makes the process valuable. With this in mind let us turn our attention away from the PDF of Gauss, at least suspected as being not a useful description of the outcome for nonsimple phenomena, and in the next chapter examine the arguments of the first scientist to recognize the existence of the long tails depicted in Figure 1.3, that being Vilfredo Pareto.

I cannot over emphasize the fact that the bell-shaped distribution is only appropriate for describing variability in simple systems, those that are linear and additive. The fact that such a model is not appropriate for research was actually recognized in a report of the National Research Council (NRC) [12]:

> Basic research is not part of a sequential, linear process, from basic research, to applied research, to development, and to application....[the] DoD should view basic research, applied research, and development as continuing activities occurring in parallel, with numerous supporting connections throughout the process.

But, unfortunately the way research in science and engineering is viewed within the DoD is strictly compartmentalized into 6.1, 6.2, 6.3 and higher. This linear view reinforces the bias towards applying the tools from the Gauss world view to the assessment of a research laboratory, to its detriment, and is critiqued by West et al. [19]. One of the purposes of the present study is to highlight the properties of STEM research that are inconsistent with linearity and to demonstrate that any reasonable measure of research quality is of necessity normal.

Beguiled by Averages:

The loss of simplicity is the reason Gauss statistics fail when applied to the behavior of people. Rose [18] points out that the psychologist Molenaar recognized that the fatal flaw in applying the Gauss view to nonsimple individuals was the self-contradictory assumption that we can understand individuals by ignoring their individuality, that is, their differences. In the Gauss view the average is everything and those pesky fluctuations that makeup individuality are all supposed to be captured in the single parameter σ. In the inverse power law (IPL) PDF variability is everything and the average is just one property of the PDF, that may or may not be important in a particular application.

But before we begin to assess the IPL distributions let us use some historic examples of just how much damage has been done by not taking into account the complexity of a problem up front and misapplying the notion of an average. Consider the game changer of introducing airplanes into the U.S. Army and the formation of the Army Air Corps (1926 to 1942). Rose [18] describes the designing of the first-ever cockpit and the use of the Gauss world view to design the cockpit to standardized dimensions:

> For the next three decades, the size and shape of the seat, the distance to the pedals and stick, the height of the windshield, even the shape of the flight helmets were all built to conform to the average dimensions of a 1926 pilot.

The problems entailed by this standardization did not become apparent until jet aircraft were introduced into the U.S. Air Force in the late 1940s with serious consequences. The pilots were not able to control the faster and more complex machines leading to frequent problems involving all manner of aircraft, which at its nadir resulted in 17 pilots crashing within the span of 24 hours [20] for no apparent reason. Rose chronicles how LT G.S. Daniels established a research program to solve the mystery surrounding the crashes. Multiple investigators had concluded that there were no mechanical failures leading up to these crashes and universally agreed that they were all due to pilot error; a judgement that Daniels found unacceptable. After conducting his own research, Daniels reached the incontrovertible conclusion, by measuring the ten dimensions that the U.S. Air Force used to characterize a pilot for each of 4023 pilots [21], that:

> There was no such thing as an average pilot. If you've designed a cockpit to fit the average pilot, you've actually designed it to fit no one.

A nonsimple process is typically described by a number of dimensions or variables that are essentially independent, that is, they are at most weakly correlated. Rose discussed a number of nonsimple phenomena that are often assumed to be simple through the kinds of questions we ask. For example we say a person is smart. But what does that mean? Do we mean their IQ is high?

Perhaps.

The IQ measure of intelligence is actually multi-dimensional. Rose provides a breakdown of the ten dimensions contributing to the IQs of two individuals, both with an IQ of 103. The inane question is: Who is smarter? The IQ score would say they are both of approximately the same average intelligence. But is that correct?

In only two of the ten dimensions being measured were the two individual scores even close. In the eight others they were all over the place: in vocabulary one was low and the other average; in knowledge one was average and the other was high; in arithmetic they were both on the low side of average. If we graphed an individual's dimension scores and joined them with straight lines as depicted in Figure 1.4, we obtain the two jagged curves whose excursions side to side bear no resemblance to one another. Rose argues that such jagged curves are typical of the results obtained for complex processes and all these individual characteristics are lost when the process is assumed to be simple and to be characterized by their average.

Reducing the variability to achieve standardization is useful in manufacturing, but in the social realm this 'discards the baby with the bath water'. It is the variability that constitutes what is important about being human. This is equally true when discussing the variability in the size of pilots, as well as in the measures of STEM innovation and research, as we subsequently show.

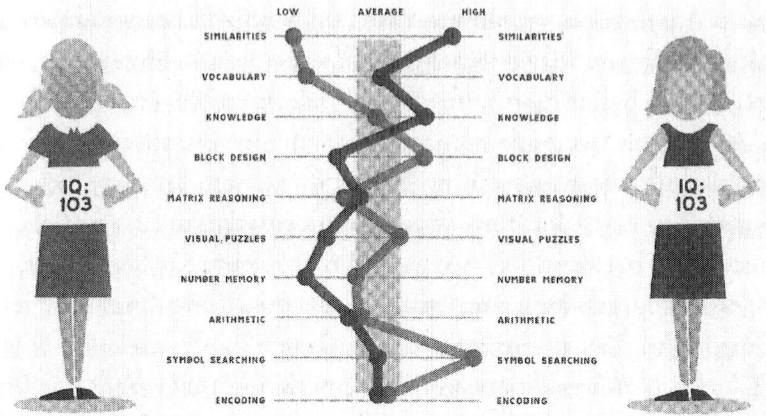

Figure 1.4. Jagged profiles over ten attributes constituting intellegence. Note the jagged dissimilar profiles for two girls haveing the same IQ. From [17] with permission.

1.5. What is to Come?

In addressing the challenges associated with the assessment of ARO research funding strategies we recognize that there are a number of issues that complicate assessments of the conduct and outcomes of basic research. For fundamental STEM research projects failure is not unexpected, research projects that fail to discover significant findings are in fact common. On the one hand, research that generates a significant discovery that can be classified as breakthrough is rare. On the other hand, research that generates no new findings is not uncommon, but may still be useful in discovering and cataloging research approaches that are not productive in specific areas of investigation and therefore ought to be avoided.

It is worth pointing out that aggregated incremental findings are often the basis for major discoveries. Research projects that produce relatively minor, incremental advances in knowledge are a necessary precursor to STEM breakthroughs. Consequently, credit for breakthrough advances need to be sought across a large number of prior research efforts. Complimentary to the incremental precursors to breakthroughs is the subsequent incremental research necessary to firmly establish wherein the breakthroughs fit within the scope of existing knowledge.

The significance of fundamental discoveries may take many years to be appreciated. History is replete with examples of discoveries whose significance was only realized many years after the findings of the research became public. For example, the importance of a particular discovery might be apparent only when our understanding of related phenomena is sufficient to comprehend the brilliance of the initial findings. Attempting to measure the 'value' of a basic research project prematurely will almost certainly provide an underestimate of its value. Einstein's equation relating mass to energy in the first decade of the twentieth century comes to mind.

Fundamental scientific discoveries may enable further advances in other related fields. In basic research, a discovery in one field may provide an insight that is applied across multiple disciplines. For example, the development of the *fast Fourier transform* in mathematics was essential to advances in signal processing that has enabled many modern innovations in information and communications technology. The importance of a discovery may be seen in research fields that are apparently far from where the discovery was made.

One implication of these features of basic research is that there is no 'average' research project—or rather, the result of an average research project is trivial. The distribution of value across research projects is highly skewed, where breakthrough discoveries are exceedingly rare but their value far outweigh the aggregated value of lesser discoveries. Normal statistical methods are not sufficiently sophisticated to account for these dynamics. This project requires a more nuanced approach to assessing the relative values of the three ARO funding support strategies discussed earlier.

I believe we have arrived at the present situation regarding the value assessment of STEM research quality through a failure to properly take into account human nature in the management of resources for conducting research, and in the facilitation of the activity of STEM investigators. To support this proposition we examine the evidence regarding the 'unfair' nature of social organizations and how this unfairness can and should influence STEM resource management. The evidence presented concerns such apparently disconnected phenomena as the distribution of income in Western society, the distribution of computer connections on the World Wide Web (WWW), the publication frequency of, and citations to, research articles, as well as, the description of other complex phenomena involving the multiple interactions of human beings. However, we begin our investigation on more familiar ground, before venturing out onto the frontier, where understanding is rare and useful ideas that have been scientifically vetted are few and far between.

It is surprising that the tools developed within the various STEM disciplines to quantify subtle variations in natural variables have not been systematically applied across those same disciplines to quantify the success of individual research activities. Nor has that scientific lens been applied to the many research programs within such organizations, nor to the government programs funding that research. Finally, such application has been made in only the most rudimentary form to the individual scientists contributing to these programs.

In a nutshell that is how we stand today.

Some will object to what I have said, assuring us that a variety of STEM disciplines have developed statistical measures of the efficacy of various indicators of research quality. In 1926 Lotka [22] graphed the fraction of the total number of investigators versus the number of papers published N as depicted in Figure 1.5. We have a great deal to say about the IPL form of the PDF and the general interpretation of the IPL index μ, shown in that figure. But for the moment to provide a perspective on the breadth of the material to be discussed we list a number of empirical IPLs that have been named after those that discovered them and which are discussed elsewhere [23]:

Pareto's Law (1896): income;
Willis' Law (1922): speciation;
Auerbach's Law (1913): city sizes;
Lotka's Law (1926): scientific publications;
Murray's Law (1927): size of limbs;
Rosen-Rammler Law (1933): fragmentation of ore;
Zipf's Law (1948): frequency of words;
Gutenberg-Richter Law (1954): size of earthquakes;
Rall's Law (1959): neuron branching;
Richardson's Law (1960): size of wars;
Price's Law (1963): scientific citations;
Goldberger-West Law (1987): size of bronchial airways.

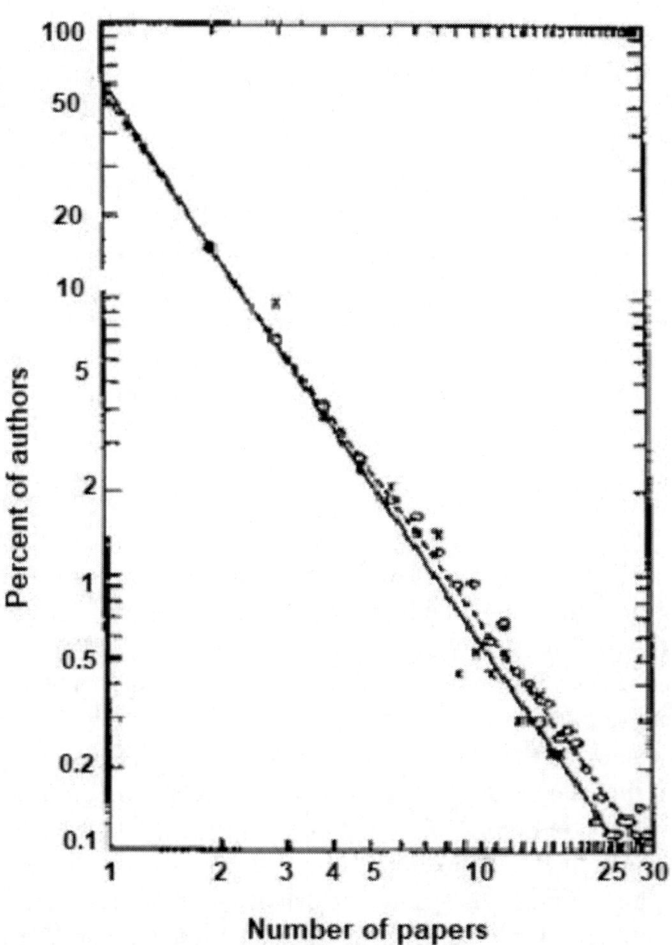

Figure 1.5. *Lotka's Law*: The number of scientists publishing exactly N papers, as a function of N. The open circles represnt data taken from the first index volume of the abridged *Philosophical Transactions of the Royal society of London* (17th and 18th centuries), the filled circles are those from the 1907-1916 decentenial index of *Chemical Abstracts*. The solid curve shows the exact inverse-square law of Lotka; the dashed line is the fit to data. All data are normalized to a basis of exactly 100 invesigators publishing a single paper [21].

However, before we can make sense of these empirical laws it is necessary to review what the traditional statistical measures tell us about such datasets. What relation do averages have with these empirical laws and where does the central limit theorem (CLT) enter the picture, if at all? It turns out that knowing how these empirical laws deviate from simple statistical theory is very important, because most of us have developed intuition about uncertainty in our nonsimple world using the concepts from simple theory. All too often, when these same concepts are applied to empirical datasets gleaned from nonsimple phenomena we encounter contradiction and paradox, not unlike the conflict that motivated the study leading to the present work. Therefore, before we apply what we know about statistics to datasets that can be used to quantify the quality of a STEM research activity, we review what distinguishes nonsimple from simple phenomena.

The empirical laws for nonsimple systems that we listed above are all IPL PDFs and the reasons for this are addressed in Chapter 2. The IPL form may not be true universally, but even when it is not true across the entire range of variation of the statistical variable, it does appear to be true asymptotically. We mentioned the CLT above. An assumption on which this mathematical argument (theorem) is based is that the second moment of the dataset being analyzed is finite, otherwise the CLT breaks down. However, in the 1920s the mathematician P. Lévy showed the existence of a generalized CLT, valid for time series having diverging second moments. A qualitative comparison of a Gauss and a Lévy PDF is given in Figure 1.6, and the asymptotic form of this new limit PDF is IPL in the tails of the Lévy distribution. It is in these tails that the Gauss and Lévy PDFs are the most dissimilar and when you think about it, this is where most catastrophic failures happen, far from the typical behavior of the process of interest [24].

The connectivity of nonsimple networks is IPL [26, 27, 28], and its influence is observed in bibleometric time series. There is a vast literature on the interpretation of IPL PDFs that has developed over the past quarter century, much of which has been applied to uncovering the cooperative mechanisms present in nonsimple adaptive networks. The reliance of the IPL PDF on the nonsimplicity of the network being examined is discussed and the relation of the nonsimplicity to the quality of the measures is examined in Chapter 2.

We apply what has been developed in the above mentioned literature to

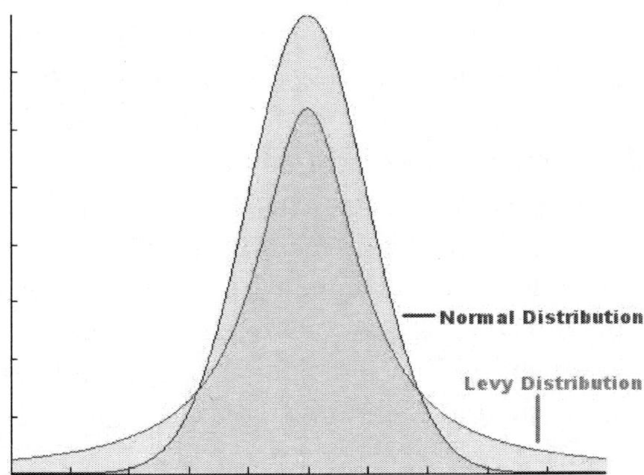

Figure 1.6. Normal PDF is labeled superimposed on a Lévy PDF. The tails of the Lévy PDF clearly has non-zero weight for large deviations from the mean of the Normal PDF. It is the tail region that produces the divergence in the second moment.

determining how the quality of a published research paper can be quantitatively measured in terms of the number of times it is cited by other research papers. The number of citations to papers across STEM disciplines are found to be IPL and consequently the average does not characterize the PDF, nor can it be used to compare the relative quality of papers without additional interpretation. If the average is not the quantity of interest, what is the proper measure of the quality of a paper? Assuming that such a measure of quality for an individual paper can be determined, the question remains concerning how this measure can be aggregated to determine the overall quality of a research program or organization.

Put more generally, the question becomes: Can the quality of STEM research itself be measured? This question was answered with a tentative yes by van Raan [29], through the use of the bibleometric analysis we discuss herein. At the next level of cooperation among investigators we quantify the quality of collaboration networks that are formed to expose, explore and solve nonsimple STEM research problems important to society in general and the military in particular.

In Chapter 3 the methods developed under the rubric of the *Science of Science* are discussed with a view to extracting patterns from the ARO datasets. Therein we present the arguments to distinguish between the number of papers published (quantity) and the number of citation each of these papers received (quality) using these patterns. In this chapter the social aspect of STEM research is emphasized and how that is manifest in the IPL form of the PDFs of the number of publications SIs have, the number of citations each of these papers garner, and the nonsimplicity of citation networks, as well as other kinds of STEM research networks. The aggregation of large datasets is used to form static measures with which to rank universities as well as nations in terms of STEM research quality.

Chapter 4 concerns itself with scientific breakthroughs in STEM research. This is done by analyzing the dynamics of STEM research networks such as SI-citation networks and SI-collaboration networks. The various SI strategies that emerge in the on-going network growth reflect the research culture the SIs in the network have adopted. Self-information is used to quantify STEM breakthroughs. The notion of an empirical paradox (EP) is used to clarify the source of a STEM breakthrough in the development of a theory within a STEM discipline.

Some general conclusions are drawn in Chapter 5 from the multiple measures developed in the earlier chapters. The theory underlying an empirical citation allometry relation (C-AR) is presented and shown to be a consequence of the statistics of nonsimple citation network. The scaling entailed by the statistics is shown to result in the observed C-AR, along with its dynamics being described by the fractional probability calculus. The exact solution for the probability of the rank-ordering is the Mittag-Leffler function and the exact PDF for the citation-frequency is a hyperbolic distribution, both of which have asymptotic IPLs.

In Chapter 6 the insight gained in the previous chapters is applied to the measures constructed by the RTI research team to quantify the quality of the ARO-supported research. The STEM research quality is characterized by three dimensions, each having six indicator measures. These eighteen measures are not used in an absolute, but in a relative sense. For each measure of the ratio SI (MURI) funding strategy to the UARC funding strategy is made for both PIs in the top and non-top universities. This provides 36 distinct relative indicators of the SI funding strategy and a similar number for the MURI funding strategy.

In summary, all evidence from the RTI study indicates that a restricted funding strategy would have significantly lowered the STEM research value of the investment to ARO. In other words, if the budget for the distributed SI funding were cut in half and instead invested in the localized funding, this would result in a disproportionate loss in research value to ARO. The patterns that emerge suggest SI awards contain distinct value different from other funding mechanisms and are an appropriate primary mechanism for funding new ideas in fundamental science with potential long-term contribution to Army capabilities.

Chapter 7 provides an overview of what has been covered. If you are less interested in the details of how we have arrived at the conclusions, and more concerned with what those conclusions are, you may want to skip the intervening five chapters and proceed directly to this chapter.

Chapter 2

Tales of Tails

At the very least a superficial understanding of the way human networks operate is needed to determine how to evaluate the quality of the STEM research carried out within laboratories, is supported by research programs, and/or conducted by STEM investigators. To achieve even this primitive level of understanding we consider how members of a group of cohorts might choose from a hypothetical set of choices, say a large set of nodes on a computer network, to which they may connect. If we assume the choices are made independently of the properties of the computer node selected, or without regard for the selections made by other members of the group, the resulting PDF of the relative frequency with which a given node is selected has the familiar bell-shaped form. In this case the selection process is completely uniform with no distinction based on personal taste, peer pressure or aesthetic judgment, resulting in a network of random links between humans and nodes, or more generally between individuals.

The bell-shaped distribution describes the probable number of links a given node has in a random network. Barabási and Albert (BA) [30] determined that random networks do not describe naturally occurring nonsimple networks. Using real-world datasets they were able to show that the number of connections between nodes on networks as diverse as the WWW or citation patterns in science deviate markedly from the bell-shaped distribution. In fact they, along with others [26, 27, 31], contemporaneously found that nonsimple networks in general have IPL PDFs, rather than the PDF of Gauss.

The IPL PDF was first observed in a systematic study of income datasets

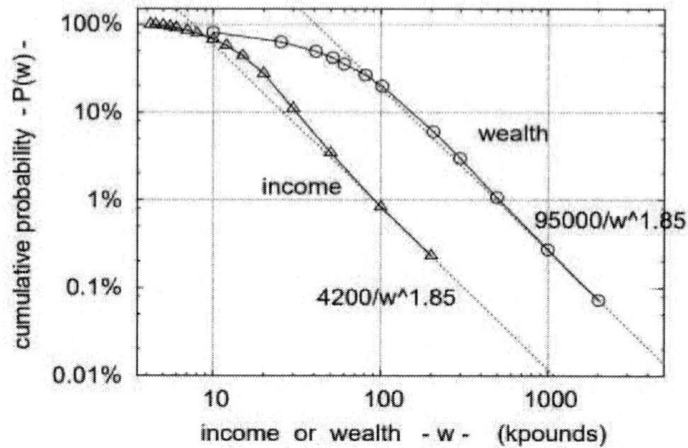

Figure 2.1. The cumulative distribution of wealth and income data for United Kingdom, downloaded from Inland Revenue, the British tax agency, is depicted. The cumulative income (1998-99) and wealth (1996) distributions on log-log graph in the United Kingdom. From [24] with permission.

in Western societies as analyzed by the engineer/economist/sociologist Marquis Vilfredo Frederico Damoso Pareto (1848-1923) in the late nineteenth century. This PDF was first recognized in the context of scientific research by the biophysicist Lotka as mentioned in the first chapter, using an IPL distribution in the number of STEM papers published in a given time interval. Subsequently, beginning with BA, scientists recognized that phenomena described by such IPLs do not possess a characteristic scale and referred to them collectively as scale-free, in keeping with the history of such distributions in social phenomena [32].

Pareto was an economics professor at the university in Lausanne, Switzerland. His analysis of income/wealth data collected from various countries led to his being the first to recognize that the distribution of income/wealth in a society was not random, but followed a consistent pattern. An exemplar of such income and wealth distributions, drawn from the United Kingdom near the turn of the 21st century, is depicted in Figure 2.1. These patterns are described by an IPL PDF, which now bears Pareto's name, and which he called: *The Law of the Unequal Distribution of Results* [33]. He referred to the inequality in his empirical

PDF more generally as a "predictable imbalance" which he was able to find in a variety of social phenomena. This imbalance is ultimately interpretable as an implicit unfairness found in nonsimple networks. We referred to this unfairness earlier, but without preamble, which may have made some readers uncomfortable.

So how do we transition from random networks with their average values, standard deviations, and implicit fairness, to nonsimple networks that are scale-free, with their IPL PDFs and explicit unfairness? [34] More importantly, how does this relate to the assessment of the STEM research quality measures of laboratories, programs and individuals?

2.1. Multiplicative Processes

Let us examine the distribution of human achievements and consider a simple mechanism that explains why such distributions have long tails, sometimes called heavy tails, such as shown in Figure 2.2. In the present chapter we discuss the distribution in the number of STEM publications per investigator, as well as the distribution of the number of citations to STEM papers as exemplars of the many IPL networks that characterize the social communities of STEM investigators. Achievement is, in general, the valued outcome of a nonsimple task. A nonsimple task or project is multiplicative and not additive because an achievement requires the successful completion of a number of separate, but tightly interconnected subtasks, in order to realize the overall nonsimple task. The failure to complete any one of the subtasks leads to an increased probability of failure of the overall task. Note that this interdependency constraint violates a central condition necessary to prove the CLT.

An exemplar of such a process that is germane to our purpose is the publication of STEM papers. Quoting the recipient of the 1956 Noble Prize in Physics, William Shockley [35]:

> It is well-known that some workers in scientific research laboratories are enormously more creative than others. If the number of scientific publications is used as a measure of productivity, it is found that some individuals create new science at a rate at least fifty times greater than others. Thus differences in rates of scientific production are much bigger than differences in the rates of preforming

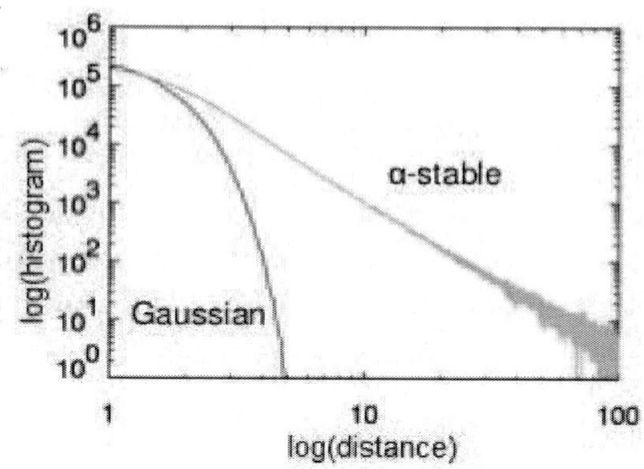

Figure 2.2. A schematic comparison is made between the bell-shaped distribution of Gauss and the IPL distribution of Pareto, obtained from appropriately chosen random walks. The vertical axis is the logarithm of the number of trajctories (histogram) and the horizontal axis is the distance travel by the walker in arbitrary units. The curve labeled $\alpha-$stable in the one-sided Lévy stable PDF which has an asymptotic IPL form.

> simpler acts, such as the rate of running the mile, or the number of words a man can speak per minute.

It was 1955, a year before he was to receive the Noble Prize in physics, and very little had been established concerning the statistics of STEM productivity when Shockley, who was then the Research Director of the Weapons Systems Evaluation Group, DoD, and on leave from Bell Labs, decided to investigate scientific productivity using as a measure the number of publications individual scientists had made. Surprisingly he was able to document that close correlation existed between the quantity of STEM productivity and the subsequent: "...achievement of eminence as a contributor to a scientific field".

Without recalling the many interesting results discovered and recorded by Shockley, we do reproduce his argument leading to the log-normal PDF of STEM productivity. The cumulative distribution is given in Figure 2.3 where 'weighted' means the fraction of a paper's authorship assigned to each individual, that is, for a paper with five coauthors, each contributor receives a weight of 0.2 of a publication. On probability paper a straight line with positive slope is a Normal PDF for the variate being graphed, which in this case is the logarithm of the weighted number of publications. Consequently, the distribution of scientific publications is empirically determined to be log-normal, which is to say that the logarithm of the number of papers published in a specified time interval is Normally distributed.

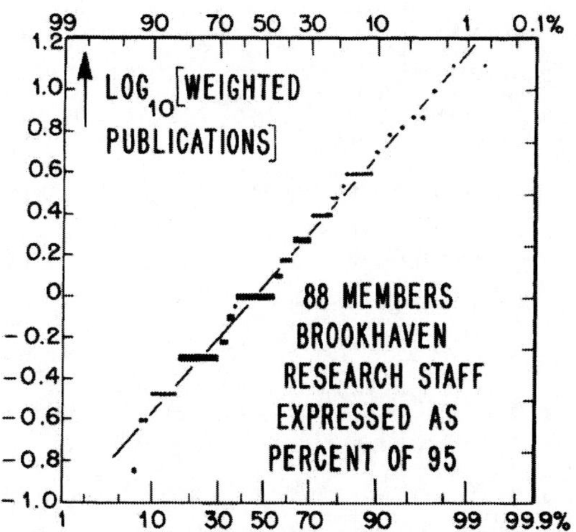

Figure 2.3. Cumulative distribution of logarithm of weighted rate of publication at Brookhaven National Lab. plotted on probabiltiy paper. The straight line indicates a log-normal PDF. From [34] with permission.

Shockley uses the factor of nearly one hundred between the least and the most prolific scientists recorded in Figure 2.3 to speculate on the reasons for the form of the PDF. One of his speculations was that in order to publish a paper an individual must possess a given number of abilities or mental factors. A partial list of abilities thought to be important to complete the subtasks necessary to publish a STEM paper is [35]:

1) ability to recognize an important scientific problem;

2) skill necessary to work productively on the problem;

3) capability of recognizing a worthwhile result while immersed in the analysis;

4) capacity to make a decision as to when to stop and write up the results;
5) ability to write simply and logically;
6) strength to profit from colleague's criticism;
7) determination to submit the paper to a journal and
8) willingness to answer referee's objections.

If we associate a probability p_j with each of these abilities $j = 1,..,8$, then to a given level of approximation, the productivity of the workforce would, based on this argument, be determined by the product of the eight probabilities:

$$P = p_1 p_2 p_3 p_4 p_5 p_6 p_7 p_8.$$

If the number of papers published by an individual out of a total of N is $n = NP$ taking the logarithm of this expression yields:

$$\log(n/N) = \sum_{j=1}^{8} \log p_j.$$

The CLT applied to this linear additive process yields a distribution for the successful publication of n papers that is Normal in $\log n$, it is a log-normal PDF.

Shockley pointed out that if one person exceeds another by 50% in each of the eight factors, that person's productivity will be a factor of 25 greater, due to its multiplicative nature. Thus, small variations in specific mental attributes can produce the large variation in the observed scientific productivity observed in a laboratory.

It is worth mentioning that a log-normal PDF is IPL at large scales with $\mu = 1$ [32]. Other less intuitive and mathematically more rigorous arguments lead to IPLs throughout the domain of the variate, but these details need not concern us here. Sornette [36] discusses in varying degrees of detail over half a dozen mechanisms that give rise to IPL PDF, not all of which are well known.

So how does the IPL nature of the statistics affect the evaluation of the quality of the STEM work force?

To answer this question we construct a familiar scenario that most STEM professionals have encountered in one form or another and participated in at some point in their career. Assume that a position has become available and a short list of candidates has been compiled. Suppose further that there are eight criteria that are being used in the evaluation of this group of candidates, all with ostensibly the same level of professional achievement. Using the above argument on the multiplicative nature of nonsimple processes, we see that if each of the criteria's probabilities for an individual is reduced by the same small factor, say 5%, then that person's total capability as perceived by the committee is reduced by 66%, or more than half, with the probability of being promoted reduced by the same amount.

On the other hand, in the additive world of Gauss the total reduction would only be 5%, in fact, this is typically how such small influences are dismissively discussed. In Pareto's world of the IPL, the world in which we live, small changes in specific attributes can result in large changes in one's career potential, but not so in the world of Gauss. This explains, in part, how an otherwise sterling career can be destroyed by a handful of minor but ill-conceived missteps.

This model of the distribution of achievement is not a Gauss additive situation, where a strong charismatic individual can oppose a promotion and cajole (convince) others on the committee into going along with him/her. This is a more subtle, multiplicative situation, where each rumor, innuendo and slur can detract in an aggressively cumulative way from a person's potential for being fully recognized. Moreover, this suggests that significantly different evaluations of individuals, or laboratories, may not be attributable to any single cause, but may be the result of an uncounted number of overlooked and often forgotten, but none the less lingering, impressions that do not contribute on their individual merits, but collectively can strongly skew a cumulative evaluation.

Here we observe that the complexity of STEM problems are described by measures with IPL PDFs , which capture the multiplicative character of the nonsimple STEM enterprise and examine the more general conditions under which we would expect to observe such behavior.

2.2. Pareto Principle

In and of itself, for most people, determining that a given dataset has an IPL PDF would be notable but not necessarily register as being fundamentally important. What makes the Pareto PDF significant are the sociological implications that

Pareto and subsequent generations of scientists were able to draw from its form. For example, Pareto identified a phenomenon that later came to be called the Pareto Principle, that being that 20% of the people owned 80% of the wealth in western countries. It actually turns out that less than 20% of the population own more than 80% of the wealth, and this imbalance between the two groups is determined by the Pareto index the IPL index μ, which is the slope of the curve of any IPL dataset on log-log graph paper. The actual numerical value of the partitioning is not important for the present discussion, since it differs from application to application; what is important is that the imbalance exists within a given area of study. However, the relative numerical value of the IPL index will be shown to be important subsequently when we compare the nonsimplicity manifest in various empirical datasets and discuss their mutual interaction.

In any event, the 80/20-rule has been determined to have application in all manner of social phenomena in which the few (20%) are vital and the many (80%) are replaceable. The phrase "vital few and trivial many" was coined by Juran [37] in the late 1940s, which he rephrased to "vital few, important many" in the enlightened 1960s. Juran [38] is also the person that attributed the mechanism to Pareto and coined the term Pareto Principle. The 80/20-rule caught the attention of project managers and other corporate administrators who now recognize that [39]:

> 20% of the people involved in any given project produce 80% of all the results;
>
> that 80% of all the interruptions come from the same 20% of the people;
>
> resolving 20% of the disruptive issues can solve 80% of the problems;
>
> that 20% of one's results are a consequence of 80% of one's effort; and on and on and on.

Much of this is recorded and discussed in a business context in Richard Koch's book: *The 80/20 Principle* [39]. This principle is a consequence of the IPL (scale-free) nature of nonsimple social networks.

IPLs are often indicative of self-organization. The phenomenon of self-organization is a consequence of the internal dynamics of nonsimple networks generating critical dynamic behavior. At criticality the microscopic dynamics no longer determine the behavior of the network but an aggregated emergent variable captures the network's collective macroscopic dynamic behavior. This aggregate variable is often an average over the microscopic degrees of freedom giving rise to a macroscopic behavior that has an IPL distribution quite unlike the dynamic behavior of the individual elements constituting the network.

The fact that IPL PDFs can be found in all manner of phenomena over all STEM disciplines, strongly suggests that this macro-behavior is independent of the detailed micro-dynamics. Rather it is a consequence of the fact that nonsimple networks often have self-organizing internal dynamics, with order apparently emerging out of disorder. In the transition from microscopic disorder to macroscopic order, as in phase transitions, networks give up their microscopic random behavior, characterized by average values and become scale-free, where the emerging macroscopic dynamic network is dominated by critical exponents [36]. It is the IPL index that determines the rate of fall-off of the IPL and captures the global character of the network; the imbalance between the rich and poor, between those that publish rarely and those that are prolific, between those whose work goes unread and unacknowledged, as well as, those that are universally read and cited by the majority of investigators in their field of study [34]. This behavior and more is implicit in Figure 2.2 where we schematically show the dramatic difference between the bell–shaped curve of Gauss and the IPL of Pareto.

It is evident from Figure 2.2 that all the action in a simple random network takes place in the vicinity of the peak of the uni-modal curve (the average value), since a random network has no preferences and is therefore symmetric. In the IPL network, on the other hand, the action is spread over essentially all the available values to which an individual has access, but in direct proportion to the number of links to that individual already present. This preference of a new member to hookup with an established member that has multiple connections, is what makes a star. A luminary in a STEM discipline (given a certain level of technical ability) is socially determined in substantially the same way as a luminary in Hollywood (given a certain level of acting ability) is determined.

From Figure 2.2 we see how unfair social phenomena are in the real world. If events happened according to Gauss, then everyone would be similarly situated with respect to the average value of the quantity being measured. If the normalized variable were a measure of income, then everyone would make approximately the same amount of money. If, rather than money, the distribution described the number of scientific publications, then the world of Gauss would have most scientists publishing the average number of papers, with a few above and a few below the average. The world of publications would be basically equitable. Applying Six Sigma to this world and introducing the appropriate constraints would have every STEM investigator publishing approximately the same number of papers. Since, by construction, there is little variability in the resultant measure this would appear to be a desirable goal.

However, that is not the world in which we live. In our world, Pareto's world of the IPL, there may not even be an average value and people may have almost none of the important variable, or they may have a great deal of it. So we find people at the level of poverty and people making millions of dollars a year and all those in between. The number of publications is no exception in this regard. The biophysicist Lotka [22], determined that the distribution in the number of scientists having a given number of publications is, in fact, IPL, just like that of the upper few percent of income earners. The difference between the two PDFs is the parameter determining the IPL fall-off; the IPL or Pareto index.

Pareto's Law was derived for the distribution of income, but there are other laws with the same IPL form: the law of Auerbach on the distribution of the sizes of cities, the law of Zipf on the relative frequency of words in language, the law of Richardson for the distribution of the magnitudes of war, the law of Lotka on the distribution in the number of papers published by scientists [40] and many others [32]. Many of these distributions stem from the implicit multiplicative nature of the underlying phenomena that make them amenable to the Pareto Principle. The essence of this principle is that a minority of input can produce a majority of the output. This is the key to understanding and assessing the quality of work produced by STEM researchers and the institutions in which they work. The Pareto Principle results in the few being important (that pesky imbalance again). In the present context, these few determine the methods, goals and direction of nonsimple STEM research teams; the STEM research direction of laboratories and in fact the overall STEM research direction of the country. The principle also maintains that the many are irrelevant in the sense that they are not the STEM leaders, but this does not imply that they are not extremely important to the ultimate success of STEM research and to the overall research enterprise. Most new results in STEM research are established by the majority, whose names we have forgotten. The 80% that refines, improvises and brings to fruition the otherwise barely intelligible ideas of the few, the 20%.

Phenomena described by IPL PDFs have a number of interesting properties, particularly when applied to social phenomena. In the case of publications, it is possible to say that most STEM researchers publish fewer than the average number of papers. It is definitely *not* the case that half the researchers publish more than average and half publish less than the average, as they would in a Gauss PDF, but rather that the vast majority of STEM analysts publish much less than the average number of papers appearing in any given time interval. Consequently, if the number of papers were the metric of quality used by management to evaluate a laboratory, most labs would not pass muster. However, this would be a distortion of the true quality of the contribution of the lab and the research being done, since this application of the metric presupposes a linear additive world view as the basis for comparison with other labs, or among members within a STEM research group. Never compare the average outcomes from nonsimple processes, such comparisons will inevitably yield distortions and not infrequently produce catastrophes.

Another IPL PDF is given by the number of citations received by published papers in any given year. The cumulative values yield the IPL depicted in Figure 2.4. In this figure, originally due to Price [41], 35% of all papers published in the sciences within a typical year have no citations; 49% of all such publications have 1 citation; 9% have 2 citations; 3% have 3 citations; 2% have 4 citations; 1% have 5 citations and 1% have 6 or more citations. The average number of citation per year in the STEM disciplines is 3.2 and the IPL distribution implies that 96% of all papers published are below average in the number of times they are cited.

Here the STEM research manager that uses the number of citations as a direct measure of the lab's quality of the research is applying a linear world view filter to the data and is not being objective about the quality of the lab's research. Most of the research published is in alignment with the 96% [42]. If the number of citations to the lab's papers is average for the discipline then that research falls in the upper 4% category. How a manager ought to use the information on publications, citations and other such criteria that have IPL statistics, for purposes of evaluation is uncertain. One thing that is certain however is that its use in evaluation process is not simple.

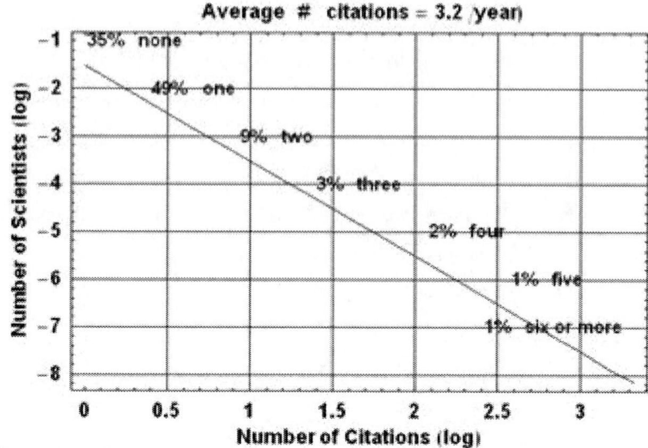

Figure 2.4 The number of STEM researchers publishing papers which receive a given number of citations within a given year is graphed on log-log graph paper. The dataset and slope for the IPL curve is from de Solla Price [40].

Consider a candidate being considered for promotion by a typical university tenure committee. The committee chair in reviewing the candidate's accomplishments points out that her research publications have the average number of citations. He goes on to say that the department wants only the best researchers to be tenured faculty and suggests that the candidate does not meet this standard. The chair has, of course, inappropriately adopted the Gauss view in which the average represent the typical and has no understanding of how STEM research is valued in the real word in general, or the IPL nature of the distribution of citations, in particular. Rather than dismissing the candidate, the candidate should be praised for producing work in the top 4% of the world's STEM research.

This failure to grasp the distinctions between what is being observed and evaluated in the very different worlds of Gauss and Pareto invariably leads to disappointment and often to disaster. Disappointment on the part of the individual who does not receive the promotion she has earned and long-term disaster for the university because other potential world-class STEM researchers will cease applying to that school for a position. If a university has the reputation that it is as difficult to get tenure there as it would be at a top-level research university then a candidate faculty ought to either apply to a top-level school in

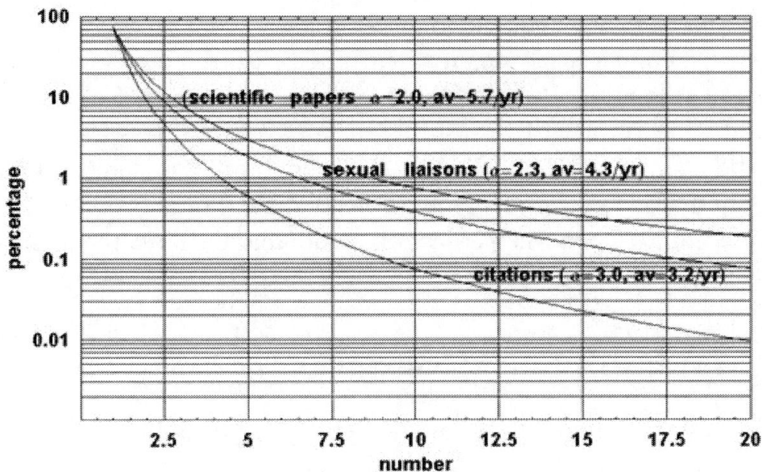

Figure 2.5. A log–linear plot of three phenomena having IPL PDFs: papers published (top), sexual laisons (middle) and citations (bottom). The data relating to science papers is from de Solla Price [75], whereas that for sexual laison is from Liljeros et al. [42].

the first place, or to a school with a more realistic self-image.

If we graph a dataset with an IPL PDF on log-linear graph paper it has the appearance given in Figure 2.5. In this figure we see that sex and STEM research are close together [43], at least in terms of their statistical properties. The IPL index of the cumulative distribution is 2.0 for the number of papers published, 2.2 for the number of sexual liaisons and 3.0 for the number of citations to a given paper. Note that the average shifts to larger values as the IPL index decreases.

Suppose a modern STEM research manager trained in Six Sigma applied that technique to either the publication or citation metric. Reducing the variability in the metric would imply eliminating 90% of the STEM analysts, leaving only those 5% above and below the average value. The rational for this action would be that the outlier researchers are deadwood, since they are either not publishing, or their published work is not being cited. The absurdity of this conclusion is unnerving. On the other hand, it ought to be expected given a Gauss world view, and it actually paraphrases a remark made by a Pentagon official a few years ago and quoted in Chapter 1.

A less Draconian response on a manager's part might be to give the 90% a substandard evaluation. But this would imply that the top 10% of the work force are judged to be average. This result is absurd as well, if less frightening, but it is none-the-less a contradiction in terms, since it implies that the best is average. Neither alternative is satisfactory and considerable thought is required to reach a notion of equity when the measures indicate that the situation is intrinsically unfair, which is demonstrably the case when the PDF is IPL [34].

2.3. Nonsimple Social Networks

A deeper understanding of IPL PDFs in the context of nonsimple social networks began with small-world theory; a theory of social interactions in which social ties can be separated into two primary kinds: strong and weak. Strong ties exist within a family and among the closest of friends, those that you call in case of emergency and contact to tell when there is a death in the family. Weak ties connect the many colleagues at work with whom you chat, but never reveal anything of substance, friends of friends, business acquaintances and most of your teachers, both past and present.

Clusters form among individuals having substantial interactions, developing closely knit groups; clusters in which everyone knows everyone else, see the regular lattice depicted in Figure 2.6. These clusters are held together by strong internal ties, but the clusters themselves are coupled to one another through weak social contacts. The weak ties provide contact from within a cluster to the outside world, see the small-world lattice depicted in Figure 2.6. It is the weak ties that are all-important for interacting with the world at large, say for getting a new job, or finding out how you are viewed professionally by the outside world.

A now classic paper by Granovetter, *The strength of weak ties* [44], explains how it is that the weak ties to strangers are much more important in getting a new job than are the stronger ties to one's family and friends. In this 'small world' the weak ties constitute the shortcuts that enable connections to be formed between one tightly clustered group to another tightly clustered group very far away. With relatively few of these weak, long-range, random connections it is possible to link any two randomly chosen individuals with a relatively short path. This short path has become known as the six-degrees of separation phenomenon [26,

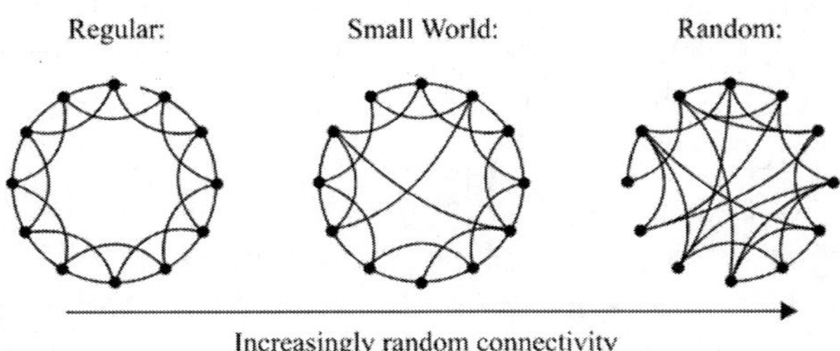

Figure 2.6. A regular graph depicts the strong coupling between nearby individuals within a group. The small world graph depicts a weak random linking of individuals that are members of distant groups. As the number of weak links increases the small world graph approaches that of a random graph.

27, 31]. Consequently, there are two basic elements necessary for the small-world model, clustering and random long-range connections.

Recent research into the study of how networks are formed and how they grow over time reveals that even the smallest preference introduced into the selection process can have remarkable effects. It has been assumed by a number of investigators that a reasonable mechanism sufficient to obtain the IPL distributions of links observed in the majority of complex phenomena is contained in the principle: *the rich get richer* [39]. In a computer network context, this principle implies that the individual with the greater number of connections attracts new links more strongly than do individuals with fewer connections, thereby providing a mechanism by which a network preferentially grows as new individuals join.

The IPL nature of nonsimple networks affords a single conceptual picture spanning scales from those in the WWW to those within any large social organization. As more people are added to an organization the number of connections between existing members depends on how many links already exist. In this way the status of the established members, those that have had the most time to institute links, grows preferentially. This mechanism has been called *preferential attachment* leading to an IPL index of $\mu = 3$ [26] for the BA model. However, the model had been developed by Yule [45] a half century earlier in the context of speciation in a somewhat more general form and further elaborated in a citation index (CI) context by Price [46], who called the CI mechanism a *cumulative advantage*. In the latter two contexts the IPL index does not have a definite value as it does in the BA model, but may be fit to accommodate datasets from different phenomena.

Thus, some members of an organization have substantially more connections than do the average, which is many more than predicted by a typical bell-shaped distribution. These are the individuals out in the tail of the IPL PDF, the gregarious individuals that seem to know everyone and to be involved in whatever is going on. In a STEM research context these are the individuals that members of the laboratory seek for technical interaction and collaboration, but with IPL indices having a variety of values [34].

Managing Nonsimplicity

The IPLs that quantify the variables of major interest to the professional STEM communities suggest that any attempt by management to equalize the reward structure within a research laboratory, such as imposing a Gauss distribution on the evaluation process, will necessarily be counterproductive and disrupt the interactive modes that would ordinarily, adaptively develop among researchers. This 'unnatural' equalization process is a consequence of the failure to recognize that like other professions, the work of STEM researchers is judged not only by managers, but also, more importantly, by professional communities of peers working both inside and outside of the organization being evaluated, say a government laboratory. The credibility of research conducted within a government lab is judged by the same STEM standards as those in the private and academic sectors outside the lab. It is critical that management recognize that the credibility of research conducted within government labs rests on the external view of the competency of its STEM investigators irrespective of locally adopted inhouse standards.

Consequently, with the 80/20-rule in mind, the data resulting from the analysis and understanding of nonsimple networks suggests that managers, evaluating the overall quality of government research labs, should concentrate on the 20% that truly make a difference and monitor in a significantly different way the 80% that do not influence the operational direction of the lab. Using this principle for guidance, the STEM managers should take cognizance of what motivates these individuals. For example, STEM individuals are typically stimulated by interesting and challenging work, but it is the top 20% that are sufficiently talented and motivated to do the basic research in the frontier areas. The other 80% can be guided to transition such fundamental research towards both long-term and short-term government needs. The two groups are complementary to one another and a successful lab requires both in order to generate innovative ideas, to carry to completion the necessary research testing of these ideas and to publish the results of the research in the appropriate journals, give talks at national and international venues to disseminate knowledge and file for the necessary patents.

Two factors that often drive entry level STEM researchers away from government labs are the lack of competitive salaries and the failure to have a reward framework in-place that acknowledges creative STEM research. An example of a creative reward for a senior STEM analyst might be a readily obtainable sabbatical at a top-level university. The lack of such rewards might be offset, in the case of more junior STEM scholars, through enhanced career advancement potential. Rewards, as well as the quality of the laboratory equipment and support facilities made available to researchers, are indications of the level of respect with which management regards the work done and even more importantly the STEM people doing the work. A progressive and well-trained management, something that is certainly desirable, is one that involves STEM researchers in the determination of the laboratory's research vision and in the selection and replacement of top-level STEM equipment, facilities and personnel.

Again taking cognizance of the 80/20-rule, it is clear that the 80% also desire high-caliber colleagues, for example, those who are well-published and whose work is cited in the peer reviewed STEM literature. The IPL nature of the number of publications and citations [42, 47] indicates the level of impact the cited research has on the STEM community and draws attention to the research activities of the government labs. Such 20% colleagues provide a stimulating environment in which to do challenging research and ensures the enhanced visibility of the lab, which increases the chances of the 80% to collaborate productively on important research with their counterparts in industry and academia.

The successful management of the government supported STEM community can only be accomplished by enabling the 20% to focus on 'unfettered' basic research, who are unreservedly supported by the 80%, who may be more engaged in the transitioning of ideas to government and social needs. Management must also bear in mind that this 20% does not always consist of the same STEM individuals. From one year to the next any individual can be one of the 20% or one of the 80%, but only if the research environment is structured in such a way as to facilitate that mobility. STEM leadership and support are important in an effective research laboratory and enabling one at the detriment of the other, in the end, serves neither.

The information scientist, science historian, and physicist, de Solla Price [46] hypothesized another regularity stemming from the curious asymmetric form of Lotka's IPL. In observing the distribution of publications of his academic peers it became clear to him (Price) that the number of individuals who dominated scientific publications within a thematic area was significantly smaller than the cohort group within that area. He encapsulated that observation into what became known as Price's law, see [48] for the mathematical reasoning that in colloquial form is expressed as:

> 50% of the work is done by the square root of the total number of people who participate in the work.

Consequently, in a STEM research organization of 10 people, all of whom work together, half the work is done by 3 people. This has significant implications for team building, or redesigning an organization of 100 into 10 teams of ten people each, so that instead of $\sqrt{100}$ or 10 people doing half the work, half the work is done by 10 groups each having $\sqrt{10}$ doing half the work, or approximately 32. This simple reorganization therefore increases the productivity of the workforce by more than a factor of three.

Other issues spring from the fact that key researchers are probably the most creative thinkers in the lab, who often times do not even realize there is a box to confine their thinking. Their innovative ideas are typically the most difficult to evaluate and those that generate them are often not the easiest personalities for others in the group to work with. When a lab faces reduction in research funding these key scientists may be the first to leave, because they can. If the lab employs 1,000 STEM workers, then considered as a group, Price's law contends that 33 do half the research. Losing 33 employees out of a thousand is not numerically significant if they are a random 33, but can be catastrophic if even a fraction of them are the creative people belonging to the 20%.

There is another way to interpret Price's law and that is that the key people create the value in the activity being considered. Whether it is in sales or STEM research the law holds, the difference is that in sales the value is quantitative and immediate, thereby producing immediate monetary rewards as feedback. The same is not true in STEM, in part because the measures of quality are not immediately evident and the feedback is consequently postponed.

2.4. S2TEM

The Science of Science applied to STEM research, or the Science of STEM (S2TEM) research, is probably a less intimidating name for the activity of applying the scientific method to determining the quality of a piece of STEM research, or to the overall quality of the research carried out within a project or institution, than is the label *Logology*. However, the latter would appear to have the wider scope, since it encompasses the study of all things related to science, whereas S2TEM would intuitively be restricted to applying the quantitative methodologies of science to the understanding of how high quality government supported STEM is achieved. A recent review of the Science of Science (SoS) [49] concluded that the value of SoS was its enabling of a more effective means of addressing social and technological problems through a deeper understanding of how to carry out successful research. Herein we freely adopt their description of science with the same kind of generalization being applied to technology, engineering and mathematics:

> ... as a complex, self-organizing, and evolving network of scholars, projects, papers and ideas...

We approach the quantification of the quality of S2TEM research from a nonsimple network perspective [50], subsequently adapting generalizations of the methodologies from the existing disciplines of scientometrics and bibleometrics in order to incorporate them into metrics that can be used to explain the patterns of STEM research production that are extracted from datasets. Herein we seek patterns in four different S2TEM contexts, as explained below. But before we begin building metrics to aid in our assessment we ought to have a clear idea of the array of available tools with which we can analyze STEM research datasets.

From the bottom up, we could start by determining quantitative patterns in the STEM research of individuals, which would include the number of papers published and the citations to those papers, in addition, the quality of the journals in which the papers are published and the reputation of those who are doing the citing of those papers. The next step up the hierarchy of patterns is the quality of the research program supporting the STEM individuals, applying the same mix of papers and citations, but now the patterns emerge at an aggregated level. The sociology of S2TEM research enters at the next level in which collaborative research networks are formed and for which the collaborative measure are more complicated than simply the numbers of papers and citations. For example, the measures could involve such things as networks of co-citation (papers consistently cited together with no common author), or of co-authorship. These and other more elaborate measures of quality will be discussed, as they particularly impact STEM research conducted within and/or supported by the U.S. government. This is what we euphemistically refer to as S2TEM.

It is also worth keeping in mind the prescient observation of V. Bush made in the decade of the 1940s regarding how scientists search through the labyrinth of publication and decide what to read, what to skim, and what to ignore [51]:

> The difficulty seems to be, not so much that we publish unduly in view of the extent and variety of present day interests, but rather that publication has been extended far beyond our present ability to make real use of the record. The summation of human experience is being expanded at a prodigious rate and the means we use for threading through the consequent maze to the momentary important item is the same as was used in the days of square-rigged ships.

Herein we propose to adapt and adopt a number of the digital measures of STEM research quality devised, extended and tested in the 75 years since Bush pointed out the shortcomings regarding the quantification of research quality and its importance with regard to one's own research interests. In doing this it becomes necessary to step beyond the traditional borders of government supported STEM research, which is the main concern of this study, to capture the insights and measures developed on the global stage and redirect their use.

S2TEM Research Quality Measures

It was not always clear whether we could measure STEM research quality writ large in the same way that we measure phenomena within a specific STEM discipline. The difficulty is, of course, the fact that research is an over arching concept that is common to all the disciplines within STEM's scope. This is both the source of STEM's strength and weakness, making the construction of quantitative measures of STEM research quality difficult. The fundamental question is: How do we measure the quality or significance of a piece of STEM research of an individual, of a project, of a program, or of an institution?

This is one of those sophistry questions that can only be answered pragmatically because what constitutes an effective measure is either a matter of definition, or is in the eye of the beholder, as is the acceptance of the definition. Moreover, it is not clear whether there is one answer or four; one for all levels of social aggregation of STEM research, or a distinct answer for each level. The way the question is posed one might expect the categories to be independent of one another, but they most certainly are not. Consequently, although some measures address the separate categories, others address the degree of correlation between categories and it is not yet clear which measures provide the more useful insights into the quality of the STEM research process as a social activity.

We adopt the point of view that certain STEM research patterns lead, more often than do other patterns, to desirable outcomes. These patterns sometimes overlap in two or more of the social categories listed in the question; other times the pattern is unique to one category. The approach we take is pragmatic, through the statistical analysis of time series. We reserve the mathematical model building until we have clear S2TEM patterns to model.

The oldest and simplest measures concern individual STEM research and consists of communicating that research and its results to other STEM investigators and compiling the responses. The purpose of the communication is to gauge whether or not an individual's time has been well spent, or squandered in pursuit of the 'Holy Grail'. STEM research is, after all is said and done, a social activity and gaining acceptance for one's work is one kind of validation for that effort. Even when the conclusions drawn from the research turn out to be wrong, as determined by additional theory and/or new experiments, the value contributed by a flawed theory is often quite useful in directing the next step along the path to knowledge away from a caloric theory of heat flow or a luminiferous aether in which to propagate light between stars in the cosmos.

In the modern era the communication of STEM research and the response of one's peers to the results of that research are done in a number of ways without face-to-face contact. Typically, a STEM researcher will organize into a coherent presentation the methods used in, and the results obtained from research, resulting in one or more manuscripts, which are submitted to peer reviewed journals. This is done to communicate the work to other STEM scholars and to open channels for future discussion. When I was a student in the 1960s manuscripts were invariably sent to various leaders in a field for comment prior to publication, thereby sidestepping the first round of journal review. However, today that step is considered to be atavistic, if not intrusive, and is almost universally omitted. Instead, once a paper is published it is tracked and the percent of investigators publishing exactly N papers in peer-reviewed journals using the same key words, in say a year, is recorded and from this distribution the significance of STEM research being done is inferred. The number of times a given paper is quoted (cited) in the published research papers of other investigators is tracked in order to quantify the reaction of the STEM community to the research. Most recently the number of views and/or downloads a paper receives are used to determine the level of interest to that work on the part of other investigators in the given STEM research area, even if the paper is not ultimately cited.

There are multiple ways a given paper may attract the attention of other investigators. The reputation of the authors is worth pointing to as a primary attractor. Being familiar with an author's previous work often prompts a colleague to read the abstract and skim the figures. If the paper is well written and the figures provocatively appealing, the casual glance becomes a download, an eventual read and perhaps even a citation. The honors a paper receives, such as being selected for 'editors choice', often directs others to its existence. We propose to analyze the various ways the co-authorship of papers influence the other measures, including the formation and evolution of collaborative networks.

Few decisions a STEM investigator makes are as important as the first one, which is the selection of the research problem to be addressed. The elements that enter into that decision are often obscure and whether there exists a pattern that successful scientists follow in making such decisions is worth investigating and if possible quantifying. Such patterns do appear to exist and to be quantifiable and to contribute to *scientific good taste*. What emerges after the fact is that others are drawn to the innovative ideas of certain talented individuals. Such ideas are typically high-risk, but when they payoff the rewards are many and varied, including the opening of subfields of research, often with a burst of publications that readily acknowledge the first investigators to explore the area and push the frontier in a new direction. On the downside, as pointed out by Fortunato et al. [49], evidence from the analysis of grant applications show that expert evaluators systematically give lower scores to truly novel STEM research proposals [52, 53, 54]. These conclusions are summarized by Bourdreau et al. [52] in the following way:

> Selecting among alternative projects is a core management task in all innovating organizations... We find that evaluators systematically give lower scores to research proposals that are closer to their own areas of expertise and to those that are highly novel. The patterns are consistent with biases associated with bounded rational evaluation of new ideas. The patterns are inconsistent with intellectual distance simply contributing "noise" or being associated with private interests of evaluators.

Chapter 3

Quality versus Quantity

As stated earlier, government archives exist containing the evaluations and recommendations for all proposals received by the various funding agencies, both those that were funded and those that were subsequently rejected. These archives contain, along with reports subsequently written by PIs, copies of published papers, awards received by researchers, etc. It therefore ought to be possible to test the assumptions made regarding the quality of various funding strategies by applying the methods of science to these datasets. So we freely adopt the description of the S2TEM acronym to be a nonsimple, self-organizing, and evolving network of scholars, projects, papers and ideas [49].

In spite of the difficulty of doing research, early on Price [41] determined that the volume of the STEM literature was growing exponentially, with a doubling time of 15 years [49] as depicted in Figure 3.1. Fortunato et al. [49] cautioned that it would be naïve to equate the growth of the STEM literature with the growth of scientific ideas. We would go a step further and suggest that it would be foolish to equate this growth with that of STEM research quality without having multiple and independent ways of measuring that quality. It would also be useful to have available the growth in the number of articles being supported by a particular service, such as the Army or Navy, to compare with the STEM community as a whole, as well as, the growth in other countries such as China, Iran, Russia, etc.

As mentioned at the end of the last chapter we start quantifying the quality of S2TEM research from the bottom up by determining quantitative patterns in

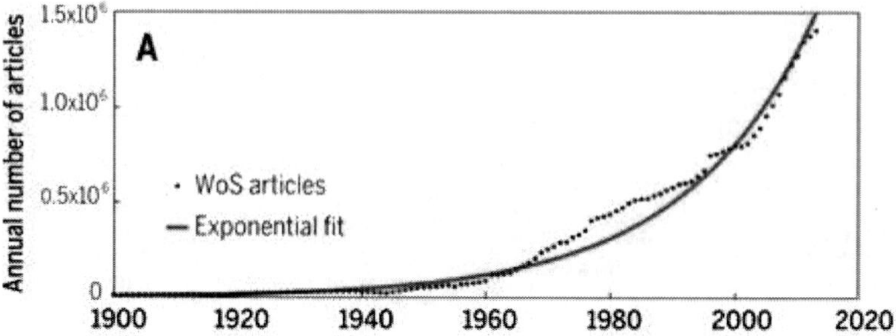

Figure 3.1. Annual production of STEM articles indexed in the web of science (WoS) database. The number of articles is given by the dots and the red curve is an exponential fit to the data. Adapted from [46].

the research of individual investigators, including the numbers of published articles and the citations to those articles. Of perhaps equal importance for initial visibility is the quality of the journals in which the articles are published and the reputations of those doing the citing. STEM research significance is not restricted to individual publications, but has to do with directly socializing the research content through individual interactions. However, it is still a matter of counting to determine the STEM stature of the individual, project or organization. How many times has one been invited to give lectures at national and international conferences and asked to be a keynote speaker? How often has one been asked to form discussion groups on the topic of their research at symposia, conferences and workshops? At what frequency do other investigators request that they share their results and insights at seminars, colloquia, or other less formal gatherings at prestigious research institutions? How many STEM journals invite them to become an associate, guest, or senior editor in, say, a year? How many STEM professional organizations have elected them to Fellowship? For example, to be a Fellow of the American Physical Society (APS), of the Institute of Electrical and Electronics Engineers (IEEE), or of the American Association for the Advancement of Science (AAAS). STEM people count things and seek stable patterns in these numbers over time. It is from these patterns that investigators uncover the underlying causes of events and if possible learn to control them. As discussed in the last chapter, Shockley was able to show from the form of the distribution in the number of scientists publishing a given number of papers that the individual productivity of scientists was sensitive to small changes in behavior and ability. Dramatic changes could result from small perturbations in distinct abilities due to the multiplicative dependence of the process of publishing a paper on these independent abilities. Therefore understanding the statistics of the quantification patterns often determines how we can interpret the underlying phenomena and perhaps formulate new policy to improve all levels of research quality in the future.

The next step up the ladder in the hierarchy of patterns is the quality of the STEM research program in which the individual engaged. Measures of the program quality are aggregates of the mix of articles, citations, and collaborations used to evaluate the STEM individual. The next level introduces the social interaction of S2TEM in which collaborative research networks are formed and for which the collaborative measures could involve such things as networks of co-citations or coauthorship. These and other more elaborate measures of quality will be discussed, as they particularly impact STEM research supported by ARO and which are part of S2TEM.

Herein we propose to adapt and adopt a number of the digital measures of research quality devised, extended and tested in the 75 years since Bush pointed out the shortcomings regarding the quantification of research quality and its importance with regard to one's own research interests. In doing this it becomes necessary to step beyond the traditional borders of ARO-supported STEM research, which is the main source of data subsequently used in Chapter 6 to capture the insights and measures developed on the global stage and redirect their use.

3.1. Number of Publications

One of the most mature and persistent measures of STEM productivity is the number of papers published in a given time interval, usually taken to be a year. As mentioned earlier, Lotka was the first scientist to notice that this measure of productivity has a pattern. In 1926 he published a paper recording his observation and that quantitative pattern is reprinted in Figure 1.5. Note that the axes in Lotka's figure are logarithms of the measured variables, so that a straight line of negative slope has the linear algebraic form in the logarithms of the variables:

$$\log Y = C - \mu \log X.$$

The distribution depicted in Figure 1.5 clearly has the form of an IPL, such that the PDF is $Y = P(N)$, which is approximately given by the number of authors publishing exactly $X = N$ papers divided by the total number of papers published:

$$P(N) = \frac{C}{N^{\mu}}, \tag{3.1}$$

where the empirical IPL index is determined from data to be $\mu \approx 2$, C is the normalization defined such that the distribution is the percent of scientists publishing exactly N papers in a given time period and is known as Lotka's law.

A half century latter in his seminal book de Sola Price [46] commenting on Lotka's law, with some degree of awe, noted that for every 100 authors producing only a single paper in a given time interval, there are 25 who produce two, 11 generating three, and so on. He went on to say that if you use the cumulative distribution, rather than the PDF, the integration gives you an IPL index of $\mu - 1 \approx 1$. In this way one could say colloquially that 1 in 5 scientists produce 5 papers or more; 1 in 10 produce at least 10 papers, etc. Again we observe the fundamental imbalance in the underlying process that is the signature of phenomena described by IPL PDFs. Subsequent measures of the PDF agree that the distribution is *heavy-tailed*, but is not necessarily IPL throughout its entire domain, as we subsequently show.

How we use the IPL PDF in the number of publications to evaluate the quality of an individual's research activity, as well as the quality of the STEM research program that is supporting the research is not obvious. This is particularly true if one elaborates on extended versions of the authorship to include two or more working in collaboration.

Before we turn our attention to these more elaborate forms of paper numberings let us examine another PDF of interest in this context. This is the distribution of time intervals between successive STEM publications by a single investigator (SI). The PDF of publication intervals is depicted in Figure 3.2 and it too is found to be IPL asymptotically [55]. This PDF indicates that it is natural to

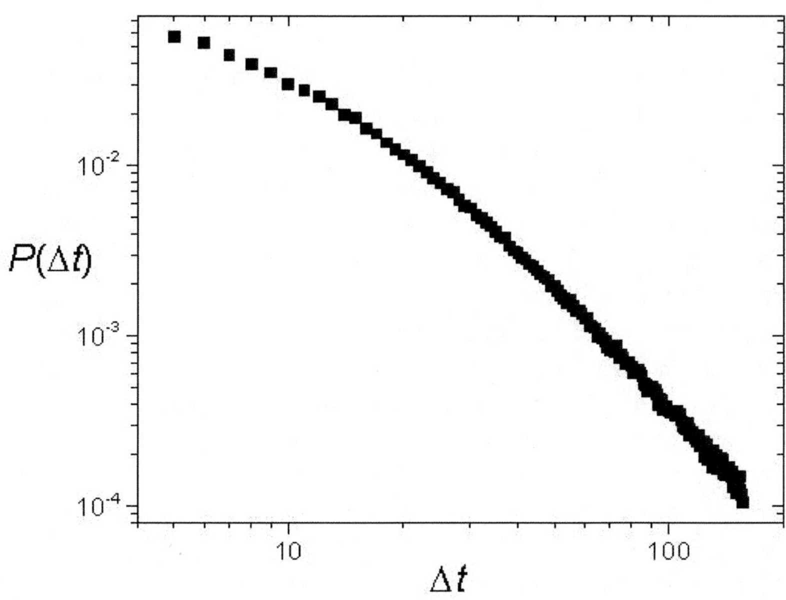

Figure 3.2. The time interval between an investigator's successive publications follows an IPL, documenting the burstiness in STEM publications. The time interval Δt in the present figure is measured in the unit of a month. From [36] with permission.

have periods of high publication activity interspersed within quiescent intervals, interpreted as the phenomenon of bursting. Therefore an individual's publication process is bursty and is reflected in the fact the inter-event time interval is an IPL where the event is the publication of a research paper. As a matter of fact, Jia et al. [55] show that the IPL form of the waiting-time PDF does not change when the data are shuffled, indicating that the publication events are a renewal process, which is to say that the successive time intervals between successive publications by an individual are statistically independent of one another.

Collaboration Patterns

One way collaboration networks may be constructed is by identifying STEM researchers as nodes of a two-dimensional lattice in which case two or more investigators are linked when they have co-authored a published paper, thereby forming a linked network of co-authorship. Newman [56] pointed out that such co-authorship networks are distinct from citation networks [41] such as those discussed in the next subsection. The publishing of a joint paper is a clear indication of significant interaction among researchers, thereby making network theory appropriate for constructing measures of patterns and the distributed nature of collaborations. An example of such a network is given in Figure 3.3 where the authorship network clearly indicates a pattern of how the collaborating STEM workers break up into disciplinary cliques.

Newman [56] selected three disciplines for study of the nature of co-authorship networks, those being biology, mathematics and physics. The biomedical network has in excess of 1.5 million authors in a five year period; the mathematics network covers a 60 year period but has orders of magnitude fewer authors; and the physics network lies between the two in terms of the number of authors. Newman comments that both Lotka and Shockley found that the distributions in the number of papers published were *heavy-tailed* which he sees in his dataset as well. The basic statistical properties for the three networks is given in the table depicted in Figure 3.4.

Significant differences in the measures of the ways investigators in different STEM disciplines do research is quantified in the table. The average number of authors on a paper differs by more than a factor of two between biology and mathematics, with physics squarely in between. The same ordering pattern appears for the three disciplines for the average number of collaborators an individual has, but now spanning a range of more than a factor of four. Newman speculates that these differences occur because biology is primarily experimental, mathematics is theoretical and physics combines theory and experiment. However, the distribution of the number of authors with exactly k collaborators is similar for the three disciplines, all having heavy-tails, as we might expect.

This difference in the statistical properties of the three disciplines suggest that a preliminary study using ARO data be done from the physics directorate because this discipline is foundational to many if not most military applications to date. This is borne out by the 10 Nobel prizes in the physical sciences over

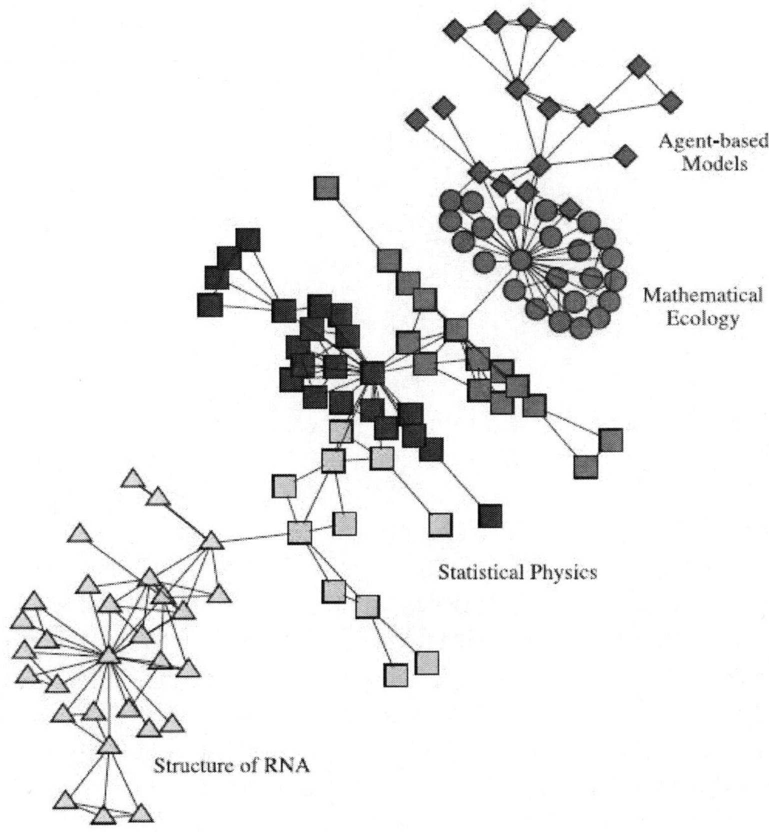

Figure 3.3. Here is depicted an example of a small co-authorship network showing collaborations among investigatorss at a private research institution. Nodes in the network represent researchers, and a line between two of them indicates they co-authored a paper during the period of study.This particular network appears to divide into a number of sub–communities, as indicated by the shapes of the nodes, and these sub–communities correspond roughly to topics of research. From Newman [55] with permission.

the past quarter century that have been awarded to investigators whose work was supported by ARO before they had attained Noble status.

	Biology	Physics	Mathematics
Number of authors	1,520,251	52,909	253,339
Number of papers	2,163,923	98,502	—
Papers per author	6.4	5.1	6.9
Authors per paper	3.75	2.53	1.45
Average collaborators	18.1	9.7	3.9
Largest component	92%	85%	82%
Average distance	4.6	5.9	7.6
Largest distance	24	20	27
Clustering coefficient	0.066	0.43	0.15
Assortativity	0.13	0.36	0.12

Figure 3.4. The statistics are, from top to bottom, the total number of authors appearing in the corresponding databases; the total number of papers appearing; the mean number of papers published by an author; the mean number of co-authors on a paper; the mean number of different individuals an author collaborated with; the largest connected group of individuals in the network; the mean vertex–vertex distance between connected individuals in the network; the largest such distance; the clustering coefficient, which is the mean probability that two co-authors will also be co-authors of one another; and the degree assortativity coefficient, which is the Pearson correlation coefficient of the degrees (i.e., number of collaborators) of adjacent vertices in the network. From Newman [55] with permission.

3.2. Number of Citations

Contrary to the romantic myth of the brilliant but socially awkward researcher working in isolation to bring an investigation to fruition, STEM research is, in point of fact, a social activity. Newton's comment about standing on the shoulders of giants" was an acknowledgement of the social interaction among investigators, with both living and dead scientists, whose contributions are necessary for the work of the individual to develop and be completed. This is not just a parsing of words, but addresses the deeper matter of the sociology of STEM research. The importance of acknowledging how prior research guides and supports contemporary investigations cannot be over-stated. Such acknowledgement not only puts new knowledge into its proper context by connecting it to what is already known, but assists in the clear communication of what new *kinds of knowledge* are being uncovered through the research.

In today's world citations constitute more than just the tipping of one's hat to those whose work has been found necessary to carry out their own research, but is a recognition of how the present work fits into the landscape of established knowledge. Whether the earlier work is background, providing motivational context for the research, addresses a mathematical challenge that had to be dealt with, clarifies an obscure technical point, or any other of the dozens of problems that typically have to be addressed or circumvented to carry out and subsequently communicate the research are all accounted for through the citation mechanism. The citations become lamp posts lighting a meandering path across a rugged landscape, revealing the knowledge gaps that the research is attempting to bridge and the blind alleys one works to avoid.

The number of times a STEM research paper is cited is a measure of the paper's luminosity and here again, as depicted in Figure 2.4, the number of investigators having a given number of citations is a heavy-tailed PDF with an IPL index $\mu > 2$. It is evident that the number of citations is often used as a proxy measure of a paper's quality, or impact it has had within a focused area of research, when in fact it could be just a indicator of passing fashion. Examples of once highly regarded but transient STEM theories that turned out to be either not as universal or useful as believed or were just wrong include: the Caloric Theory of Heat, Eugenics, Catastrophe Theory, Spontaneous generation, Psychosurgery (lobotomy), Heliocentrism of Copernicus, and Cold Fusion, to name a few in no particular order of importance, nor time in history. Consequently, any patterns, over and above the citation PDF, are useful in further assessing the quality of STEM research.

In thinking of the citation process as a unidirectional linking of papers within a network, it is evident that such forward interactions in time of one paper on subsequent papers reveals an opportunity for clustering. Such clustering can take the form of two papers always being cited together, their being co-cited. An additional paper can be added to the two, resulting in a three co-citation. Van Raan [57] discovered that like many other nonsimple phenomena STEM knowledge manifests clustering in its growth. Using the 6 million cited papers from the year 1984 and restricting the analysis to the 70,000 most highly cited papers and the 10,000 clusters they form, results in the IPL distributions of two and three co-citations clusters depicted in Figure 3.5.

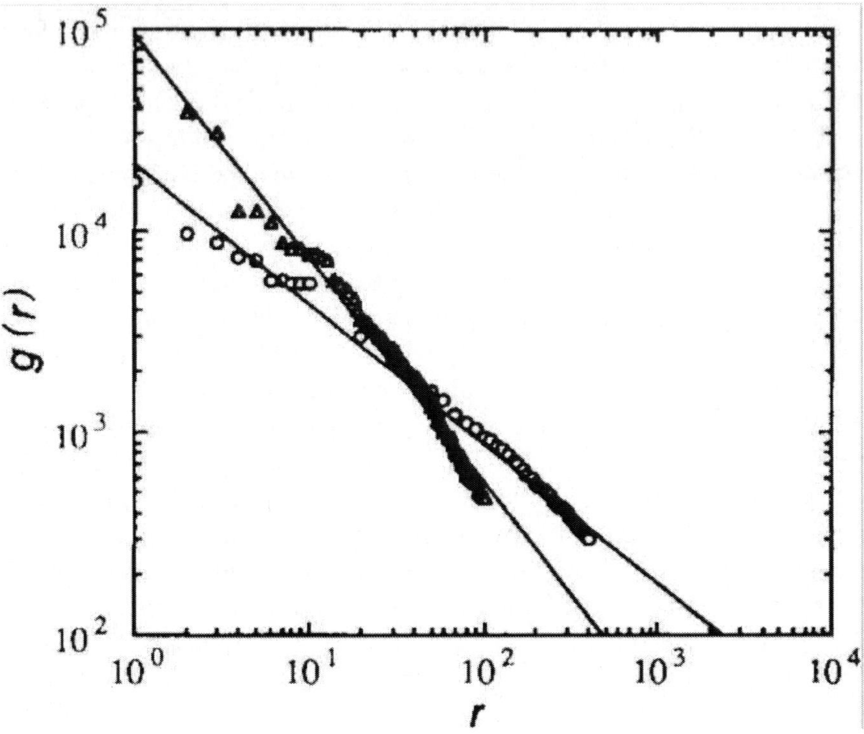

Figure 3.5. Size–rank distribution $g(r) \propto r^{-\alpha}$ is plotted versus the size of the co-citation cluster r. Circles, are for the IPL part of the (total) 1,371 two-citation clusters with fitted exponent $\alpha = 0.71$; triangles are for the IPL part of the (total) 179 three-citation clusters with fitted exponent of $\alpha = 1.09$. From [56] with permission.

Compare the IPL index obtained for the two and three co-citations clusters depicted in Figure 3.5 with the IPL index originally obtained by Price [41] to be in the interval $2.5 \leq \alpha \leq 3$. Such comparisons indicate a non-monotonic ordering in the IPL index with respect to cluster number. Van Raan [57] was able to interpret this clustering process in terms of fractal dimensions, which may have some value in generalizing to the notion of a fractal citation network and providing a quality measure of a collaboration network.

Another pattern that is sought is in the citation growth of newly published papers and whether or not things such as the early rate of growth in the number of citations can be used to determine if a particular piece of research is disruptive. How long after its publication does it become clear that a particular research result is breakthrough or not? Is there a detectable pattern to the number of citations, or in the reputations of those doing the citing over time that would warrant a shifting of resources to facilitate the development of the research area? On the flip side are there citation patterns in well-funded areas of research that would suggest that the scarce resources for the support of high-risk research be redirected to other more speculative research ideas? These and other similar questions are asked and routinely, if not systematically, answered by PMs every day.

SI Research Quality

Traditionally, the metric for the quality of a SI's research has been based on citations to that person's papers. The total number of papers published, along with the total number of citations to all papers are the first two metrics that come to mind. These are followed by the average number of citations per paper; the number of papers with citations above some threshold value, and the percentage of publications that are highly cited. These measures have been superseded by one that combines the numbers of papers published with the citations to those papers. This is the $h-$index, which equals h if a person has h papers with at least h citations and is an attempt to bring the measures of quantity and popularity into a single measure of quality.

Redner [58] did a statistical analysis of this measure to test a scaling result assumed in the original work of Hirsch [59] in which this index was introduced and developed. The assumed scaling was that the square root of the total number of citations to an individual's publications $\sqrt{N}$, scales linearly with that person's $h-$index. Figure 3.6 is one result of Redner's analysis of data for 255 physicists to obtain a PDF that clearly peaks in the vicinity of

$$s \equiv \sqrt{N}/2h = 1,$$

suggesting that $\sqrt{N} \approx 2h$, but without explanation (interpretation) as to why the equivalence ought to exist. A second conclusion Redner drew was based on the

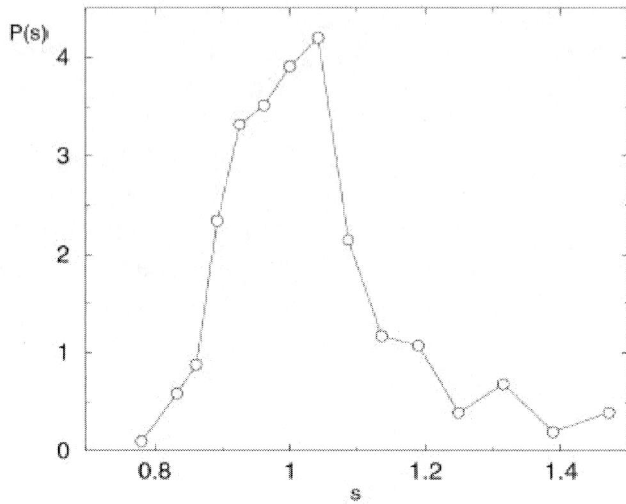

Figure 3.6. Plot of the PDF $P(s)$ that an individual is characterized by a value $s = N^{1/2}/2h$, where N is the total number of citations. From [57] with permission.

dispersion of data points around the diagonal $\sqrt{N} = 2h$. He contended that research excellence is suggested by the condition $N > 4h^2$. This conjecture seems reasonable since excellence would not share the statistical behavior of the majority of competent but not extraordinary investigators. Redner also pointed out that this is a sufficient not a necessary condition for excellence. The condition is not necessary because of the nonsimplicity of the dynamics and patterns in both publishing and citations provide alternative ways to excel.

It is well known that the production of publications and citations vary dramatically across STEM disciplines, as well as, with the stage of a scientist's career given the growth in the number of papers published. Consequently one drawback of the $h-$index is that it cannot be used with confidence for across-discipline comparisons. Another weakness is the inability to compare the prestige of an individual early in their career to one who is farther along. A substantial number of proposed corrections to these shortcomings have been made, a half dozen of which are reviewed in [50] along with their various rationales. One correction takes into account the bias introduced by treating all papers cited in a h–measure equally and takes into account the importance of having a number of highly cited papers when determining a STEM researcher's prestige.

Hirsch conjectured that a researcher could produce papers of similar quality at a steady rate over their career. Say that rate is m and although this rate can vary substantially from individual to individual, it is useful to have in mind a rule of thumb posited by Hirsch, who concludes after inspecting the citation records of many physicists and applying his theory:

> (i) A value $m \approx 1$ (i.e., an h index of 20 after 20 years of scientific activity), characterizes a successful scientist.
> (ii) A value $m \approx 2$ (i.e., an h index of 40 after 20 years of scientific activity) characterizes outstanding scientists, likely to be found only at top universities or major research laboratories.
> (iii) A value $m \approx 3$ or higher (i.e., an h index of 60 after 20 years, or 90 after 30 years), characterizes a truly unique individual.

The efficacy of citations as such a proxy measure for research quality has been called into question. For example, King et al. [60] find that over the last two hundred and fifty years, ending in 2011, 10% of the cites to research papers are self-citations and that men cite their own papers 56% more frequently than do women. Interestingly, women are also 10% more likely than men to *not* cite their own previous work at all. They provide preliminary explanations for these observations, which I think reveals more about the authors than it does the phenomena being explained.

Many STEM researchers and most non-STEM individuals believe the myth that investigators do their best research in their youth, even though this is not what the data show. Sinatra et al. [61] found that the greatest-impact research in an investigator's career is randomly distribution over their entire body of work. The probability of the greatest-impact could appear in the first work, in mid-career, or in the last paper published and also appears to be independent of the STEM discipline, with different lengths of career, at different times and in different modes of research, whether working alone or in teams. They use this empirical result to construct a quantitative model to systematically untangle the roles of productivity and luck in a STEM researcher's career.

3.3. Aggregate Measures

Decisions made by ARO's STEM PMs, as well as by senior Army leadership, regarding where to invest scarce resources often take into consideration the ranking of the research university hosting a SI's proposal. Unless it is consciously removed a pre-read SI proposal from Harvard or MIT receives greater weight than a similarly pre-read proposal from any little-known university from outside the urban centers in which the top universities are found. It is therefore reasonable to determine how an university's research rank is ascertained in order to mitigate if not entirely remove this confirmation bias from the decision-making process. Does the fact that a university is of rank one, ten or one hundred on some rank-ordered list really make any difference in the quality measures of the research performed? Of course, it makes a difference in how the university is viewed, but more importantly for our purposes here is the question: Is the rank order of a university a reliable ongoing measure of the relative quality of the STEM research being done there?

There is a great deal of hyperbole surrounding the research rank of a university, particularly since they have become big business. This situation often elicits the question: What fraction of the hype is based on quantifiable information regarding the research carried out by individual faculty or research collectives within the institution and what fraction is generated by innovation on the part of the public relations department? All too often the STEM ranking of a school is viewed in much the same way as the league rankings of their football or basketball teams. The latter, overly simple view, was critiqued by van Raan [62], who emphasized the difficulties in making bibleometric assessments of the research performance of institutions such as universities. One constraint is that the nonsimple world of STEM research is compressed into what appears within the categories of output and impact. The output is the number of STEM publications, books, letters, papers, essays, etc. appearing in a given time interval and the impact is the number of citations over a time interval to a published work, as well as, the honors garnered by those generating the output.

Research Wealth of Nations

We can, of course, start at a higher level of social aggregation than the university and compare the research status of nations. As King [64] pointed out a nation's

STEM standing is vital for policy makers in the public and private sectors that must decide STEM priorities and funding. Using the measures of output and impact, given by the number of publications and citations, from Thompson ISI, King determined the ratio of the total number of STEM papers and reviews from 1993 to 2002 for the United States and the EU15 [1] in Figure 3.7. Note that the relative number of publications had leveled off by 1999, but the relative number of citations continued to rise. It is probably time to repeat this analysis using the last quarter century as a database.

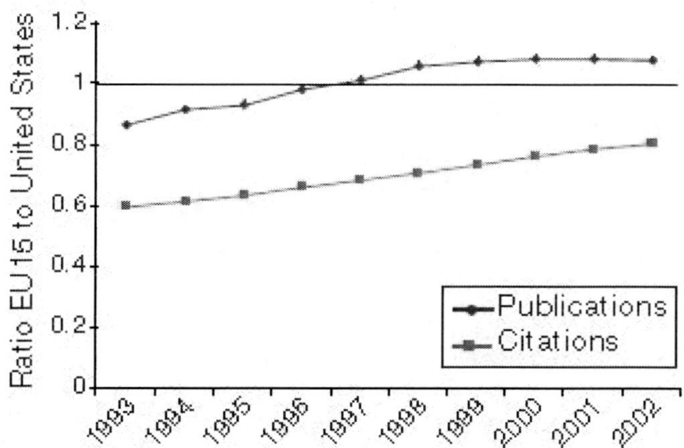

Figure 3.7. Comparing ratios of STEM publications and citations of the 15 European Union countries in the comparator group (EU15) to the United States on ISI databases in 1993–2002. The EU15 total contains some duplication because of papers jointly authored between countries in the EU group. Counts for papers and citations are totals for country (or group) for the stated year. From [40] with permission.

Note that we have labeled the STEM productivity of an individual or group with the more neutral term 'output' (e.g., the number of publications). The STEM quality of a piece of research has similarly been replaced by the less

[1]The EU15 consists of the following countries: Austria, Belgium, Denmark, Finland, France, Germany, Greece, Italy, Luxembourg, the Netherlands, Poland, Portugal, Sweden, Switzerland, and the United Kingdom.

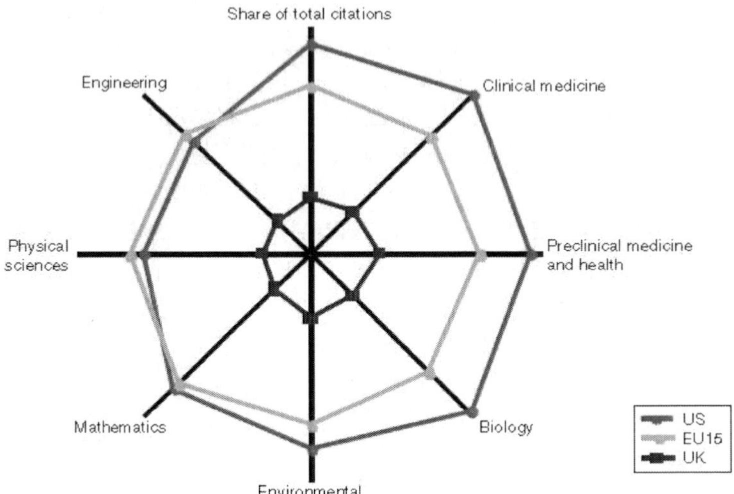

Figure 3.8. National strengths in different STEM disciplines. Plot shows research footprints for the G8 nations excluding the United States, based on the national share of citations in each of seven disciplines and overall percentage share of citations. The distance from the origin to the data point is proportional to the citation share. The medical and life sciences are shown to the right, mathematics and physical sciences to the left. From [40] with permission.

subjective term 'impact' (e.g., the number of citations a publication receives). Moed et al. [63] impugn the use of the term quality in a STEM research context, in part, because of how difficult it is to operationalize the general concept.

Another way to compare the research status of nations that does not rely on a single number, such as the citation index (CI), but breaks out the major contributions to the STEM research enterprise is depicted in Figure 3.8. The comparison of the United States to the EU15 in this figure shows that the United States, during the last decade of the 20th century, had a greater disciplinary footprint than the EU15, and it was largely skewed towards the life sciences, as indicated by the outer-most curve in the figure. Note that the relative strengths and weaknesses of STEM disciplines appear as asymmetries in the total number of citations within that discipline. The UK research footprint can be seen as a scaled version of the US and seems to have the same priorities. The EU15 research footprint, on the other hand, is clearly more symmetric, being slightly stronger than the US in the physical sciences and engineering, but weaker in the life and medical sciences. Again we emphasize that this was the situation at the onset of the 21st century and is probably quite different today.

Due to the ever increasing global emergence of STEM research collaboration networks [65] it would be of value to construct a diagram similar to that of Figure 3.8 for the adversaries of the United States, in particular for China and Russia. Adams, the Director of Research Evaluation for Evidence part of Thomson Reuters, observed [65]:

> If the science superpowers are to avoid being left behind, they will need to step out of their comfort zones to keep up with the dynamism of the new players in the shifting landscape.

Ranking Institutions

It would seem reasonable that the utility of this snapshot comparison of nations on the international stage could be reasonably adopted for the comparison of STEM research organizations on a national stage, such as for universities. But to the best of the author's knowledge this has not been done. So we turn to methods that have been applied to determining the rank ordering of US universities, including generalizations of the h-index beyond that of the individual STEM investigator to research organizations.

The h-index increases with the number of papers published N and therefore there is the danger of misinterpretation because a high h not only reflects the quality of the research but also the number of papers. Unexpectedly in the generalization considered by Molinari and Molinari [66] a scaling relation between h and N is revealed:

$$\overline{h} = h_m N^{\beta}. \tag{3.1}$$

Note that this h-index is the product of what they labeled an impact index h_m, and a power law in the total number of papers published N. It went unnoticed by these and subsequent authors over the past decade that Eq.(3.1) has the form of an allometry relation (AR) in which the average functionality $\overline{h}$ of a complex citation network increases nonlinearly with the size N of the network. Here for $N > 200$ the empirical parameter h_m is a size-independent constant, as is the non-integer exponent β. The exponent was found to have the empirical value

$\beta \approx 0.4$ for all the universities analyzed as depicted in Figure 3.9. The number-independent impact index can be used to rank order the various sources of data (universities, laboratories or journals). Note further that their analysis predicts that Harvard and Caltech are comparable STEM research institutions as is seen by comparing the boxes (Harvard) and crosses (Caltech) in the figure.

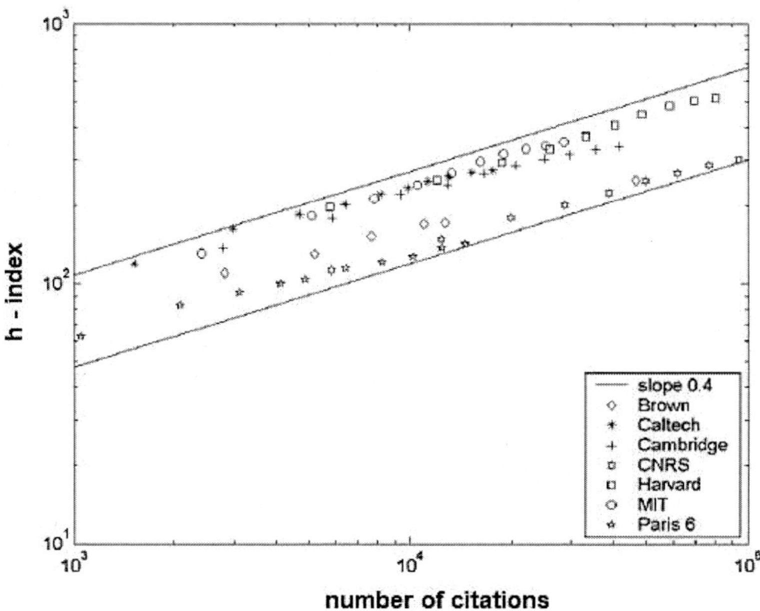

Figure 3.9. The $h-$index is depicted as a function of population size for various international STEM research institutions; note that for the large numbers of papers considered the power law exponent is approximately 0.4, which is asserted to be universal. The vertical shift between institutions is an indication of their respective visibility. The sample sizes were built by assembling papers published in even years starting in 1980 up to 1998. The two continuous lines with slope 0.4 are added on the figure to aid the eye in determining the general growth rate, but are not upper or lower bounds. Indeed, most universities would fall under the lower straight line. Figure from [58] with permission.

We point out here that each citation in the above analysis carries the same weight. A graduate student citing a paper according to this measure has the same weight as that of a Nobel Laureate, but we know that is not the case in the real world. This limitation of the measure is mitigated by putting a researcher's work in a context where the reputation of those quoting the work is taken into account. This is done when we consider the STEM network to which the investigator and the work belong, which we show how to do in the next chapter.

Another strategy to characterize research institutions using citations was made by Chatterjee et al. [67], only they were not seeking to rank-order the institutions. We know from our earlier discussion that the publication rate was determined by Shockley to be dictated by a lognormal distribution, but that the tail of the citation distribution of individual publications, as well as the number of papers, decay as an IPL in the independent variable [47]. The consensus is that both are true. Chatterjee et al. use this result to establish, for the first time:

> ...that irrespective of the institution's scientific productivity, ranking and research impact, the probability $P(c)$ that the number of citations c received by a publication is a broad distribution with an universal functional form.

The universality of the form of the PDF depicted in Figure 3.10 argues for the underlying mechanisms that produce the PDF to be irrelevant and the PDF's functional form to be the result of criticality in the underlying social dynamics producing a citation and the citing of a paper to be considered an event. The particular curve we have elected to show in Figure 3.10 contains citation data covering a five year interval, but the same result is obtained on a year by year basis, with suitably adjusted parameters. For each year of collected citation data for an academic institution the distribution $P(c)$ was observed to be quite broad. Chatterjee et al. [67] rescaled the value of citations for each year by the average number of citations per publication $\langle c \rangle$, and plotted the quantity $\langle c \rangle P(c)$ versus $c/\langle c \rangle$ on log-log graph paper, in which case all the PDFs from the various institutions collapse onto the universal curve shown, independently of the variability in output of the different institutions.

Another useful parameter for the research institutions is the k−index, which is interpreted to give the top cited $(1-k)$-fraction of papers, which have k-fraction of citations. The empirical value obtained in the study is $k = 0.75 \pm 0.04$ suggesting that the top 25% of the articles garner 75% of the total citations for an institution. Recall the discussion of the 80/20-rule for Pareto IPL distributions

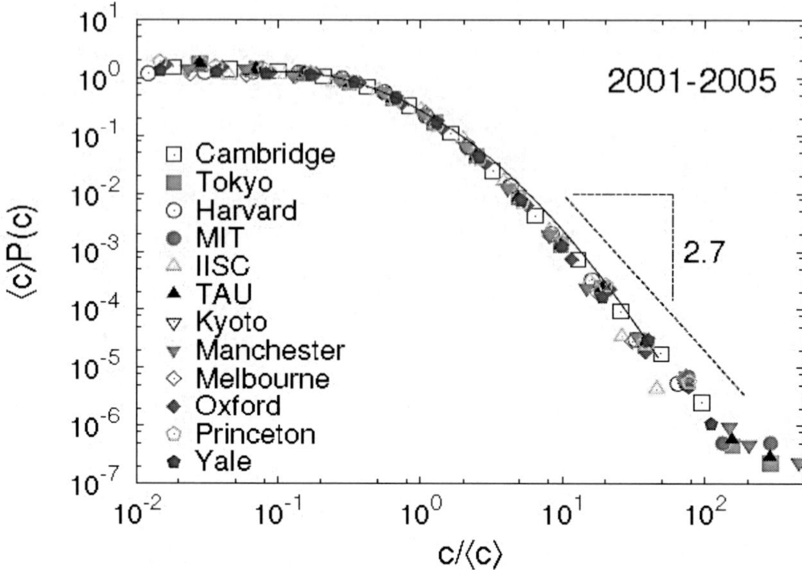

Figure 3.10. PDFs given by $P(c)$ of citations c rescaled by average number of citations $\langle c \rangle$ to publications from the period 2001-2005 for several academic institutions. Most of the range of the data fit well to a lognormal function with $\mu = -0.60 \pm 0.04$, $\sigma = 1.28 \pm 0.02$, but the highest number of citations are fit to an inverse power law, $c^{-\alpha}$, with index $\alpha = 2.7 \pm 0.3$. Obtained from [19] with permission.

in which 80% of a network's function is accomplished by 20% of the network's members.

One indication that these general results are of broad STEM interest, but perhaps of less interest for specific S2TEM applications, are contained in the results obtained by Néda et al. [68] depicted in Figure 3.11. In this latter figure we see the same universal behavior observed in Figure 3.10, but these latter data are fit by a hyperbolic universal form, rather than the combined lognormal and IPL PDFs. We note here that these authors named this distribution the Tsallis-Pareto distribution, but the form has a long history established in the mathematics literature as the hyperbolic distribution [69], which is the name we use here.

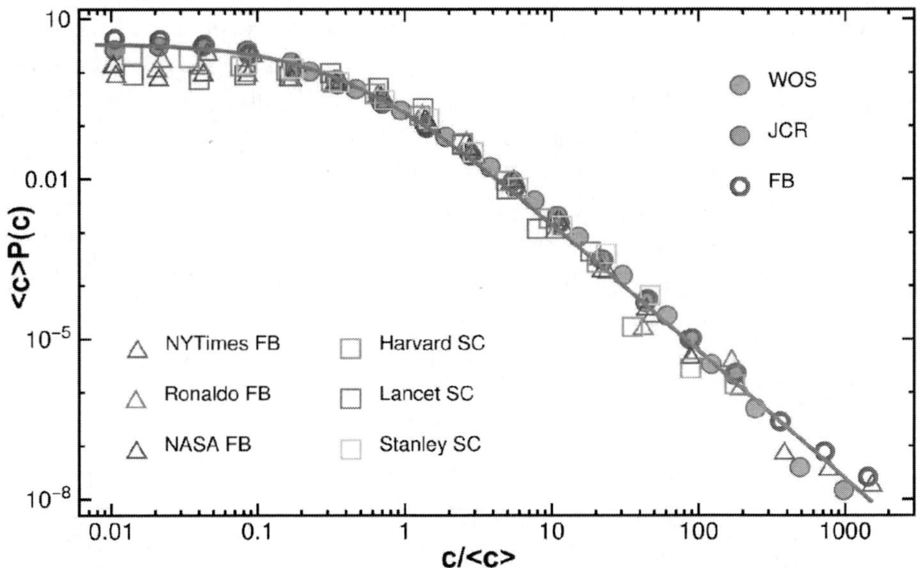

Figure 3.11. Rescaled distribution of the citation (share) numbers: $P(c) = f(x)$ is the PDF for one paper (post) to have $x = c$ citations (shares). We present the $\langle c \rangle P(c)$ value as a function of $c/\langle c \rangle$ ($\langle c \rangle$ the mean value, or first moment of the PDF). For high citation number a clear IPL trend is visible. Different symbols are for different datasets as illustrated in the legend. The considered datasets are described in [63]. The entire curve can be well-fit with a hyperbolic distribution with an IPL index $\mu = 2.4$ and $\langle c \rangle = 1$. Obtained from [63] with permission.

Another difference between the last two figures is that the latter authors have included the distribution of Facebook shares to compare with STEM citations in the data analysis. Here again we emphasize that the universal form of the citation PDF emerges from the underling nonsimplicity typical of critical phenomena. STEM and Facebook datasets each show the same popularity pattern described by a hyperbolic PDF that can be understood using a dynamic model [70].

The universality form of the hyperbolic PD suggested to Néda et al. [68] that Western society acts responsibly and in a selective manner in retransmitting information. We speculate that they reached this interpretation based on the way STEM researchers like to think about how information is transferred within a citation network in STEM research. We believe that their results do not justify this interpretation and instead are a consequence of the efficiency of information transfer in dynamic networks undergoing criticality. However, given the universal form of the PDF, we know that we cannot reconstruct a network's micro-dynamics from the emergent macro-dynamics since the emergent behavior is a consequence of criticality and cannot be unambiguously tied to any details. Consequently, the fact that STEM citations and Facebook shares have the same universal form of a hyperbolic PDF does not imply that they share a common network dynamic origin, but they do share the critical behavior of nonsimplicity.

Chapter 4

Extremes

When talking about research and researchers at dinner parties the conversation often turns to those that are famous and have made major contributions to their fields of inquiry. The names of men that typically come up are Einstein (Noble in Physics for photoelectric effect) and Barnard (first heart transplant) and the names of women are typically Nightingale (introduced statistics into healthcare) and Curie (Nobel in Physics for research in radioactivity), or others, depending on the guest list. These are people that almost everyone knows about because of their media fame: newspaper stories, magazine articles, Wikipedia entries, blogs, etc., not necessarily because of their contributions to STEM and/or to society. They were and are the best of the best, but what lead them to do the work that precipitated such world-wide recognition?

It is not the popularity of these and other high-achieving STEM researchers that we are interested in here. Rather it is the reason, or reasons, they selected the high-risk research paths they took and why they stayed the course. Was the extreme research path a matter of STEM good taste, like buying a work of art from an unknown artist, because they saw what others did not see (yet)? Or was it good fortune and they just happened to like the painting before the artist was recognized as a savant by the art world as a whole? Of course it could also have been a blend of the two, or any of a wide collection of other reasonable conjectures. The question of interest is whether the quality of the STEM research is quantifiable, whether referring to the path taken, or the problem selected. Or perhaps more clearly stated, is the path predictable? At a certain time and given a certain history, is it possible to predict what will be the next breakthrough in a specific area of STEM research, if only a person with the requisite skills has access to the resources they need to design and carry the research program to completion?

The answers to these questions are important to the STEM research institution where investigators reside, whether it is in the tradition of the academy, or a national research lab, such as the Army Research Laboratory (ARL). First of all the answers are important because the reputation of an institution, which is the only credit an organization has in the marketplace of ideas, rises and falls with the recent quality of the research done by its members and how that quality is promoted in the STEM community and popular press. Secondly, the institution taxes the research budget through overhead charges on incoming research grants, money it can use to attract high-potential young investigators, renovate aging laboratory equipment, and support staff researchers who are between research grants. Having available a quantitative method for determining a viable extreme research strategy would greatly assist in the evaluation of research programs being developed by senior STEM researchers within the organization and reduce the risk of investment in innovative through untested ideas.

The question regarding the quality of the research path is equally, if not more, important to answer for the STEM PMs at such research funding organizations as the ARO. A PM's duties include the determination of the potential viability of proposed research prior to providing resources to support that research. Once the research is funded the PM must assess the timely progress of that research toward the stated STEM goals outlined in the proposal and elaborated on in the grant. Moreover the stability of the funding of the research over time is determined in large part by the confidence the PM has in the PI, as well as by the reception the published research receives from the STEM research community as a whole and the feedback they provide. PMs must also determine whether the research programs they put together are at the extreme of what is possible, while being able to quantify the likelihood of the next high-risk proposal admitted to their program striking oil or of being a dry hole.

4.1. Single Investigators (SIs)

In previous chapters we argued that the primary measures that have been developed to quantify the quality of research are based on measures that use the numbers of citations received and papers published as input datasets. This is also true of the SI citation networks (SICNs) that have been analyzed over the past two decades, as well as the collaborative SI networks (CSINs) over the same period. The former establish patterns of investigators who quote and use one another's research to further their own investigations. The latter establish patterns of investigators who actively collaborate with others outside, as well as those within, their own institutions on one or more projects to advance their STEM research.

However, it is becoming ever more apparent that the emergent properties of such SICNs and CSINs may quantify behavior that could not have been observed, much less predicted, from a reductionist assessment of the underling dynamics of citation-linkings of papers, or the collaborative biases of SI's in various disciplines across time. These nonsimple dynamics are made up of the myriad of interdependent decisions the SI's make in the initial process of selecting a research topic and the subsequent publishing of one or more papers based on their research. The publishing process discussed in Chapter 3 is here extended to include who to cite, as well as what to cite, to put the SI's research into a broader STEM context. But long before writing the paper they must choose the STEM research problem to investigate.

The least nonsimple kind of SICN is a network where the nodes are papers and the links are citations; the authors of one paper quoting an earlier paper thereby forming a unidirectional link. We find that this in itself reveals things about the quality of the STEM research being done that only becomes apparent at the network level. The clustering of citations to form disciplines emerge at first, followed by the more tightly clustered formation of sub-disciplines within the disciplines. The patterns that emerge provide insight into how SIs view their research culture and what they think is important within that culture.

It should be evident that we can restrict the definition of a SICN using any distinguishing property of the papers being cited that we wish. For example, a single SI's research work may be quantified by constructing a SICN consisting of all papers in which that work is cited. This latter network provides the citation information on which all the measures discussed in Chapter 3 were constructed. However, the information that can be extracted from a CSIN is quite different from that uncovered in the earlier discussion as we subsequently discuss.

Research Strategy

Foster et al. [71] answered the question regarding the possibility of quantifying the quality of SI research strategy in the affirmative. They hypothesized that the choice of research problem is made as the result of an "essential tension" between a drive toward risky innovation and a conflicting professional demand for productivity. This is not unlike the tension every product manufacturer faces when presented with an idea for a new product, that being, whether to stay with the established brand that has made money for them over the years and not pursue the new idea, or take the risk of investing resources in a new product line that may ultimately fail. It is a tension that develops in every realm of human activity involving innovation, where the new and bright replaces the familiar that has apparently lost its luster over time. This is one version of the well-known innovator's dilemma.

They [71] approached the solution to the innovator's dilemma in this form using nonsimple networks on which they define the research strategy depicted in Figure 4.1. The figure depicts the evolving state of the scientific knowledge of chemistry, involving the interaction within and between two knowledge clusters. Note that the depicted network is not like the interactive SICNs discussed in Chapter 3, where the citation links between papers determine the static nature of an established network. It instead refers to the dynamic behavior of a SICN over time. As more papers are written and the authors decide which of the papers within the existing network they will cite in their own work, the properties of the SICN, such as the formation and development of clusters, or their collapse, emerges. The form taken by the aggregate of these decisions over time constitute the SIs' research strategy, and when the aggregate strategy appears formless it is labeled random. The emerging strategic property of the growing SICN, whatever its form, is unconsciously dictated by the STEM research culture into which the SI has opted.

The scientific discipline on which they [71] elected to test their ideas was chemistry, using abstracts from the National Library of Medicine's (NLM's) MEDLINE collection, focusing on papers published in the years 1983–2008

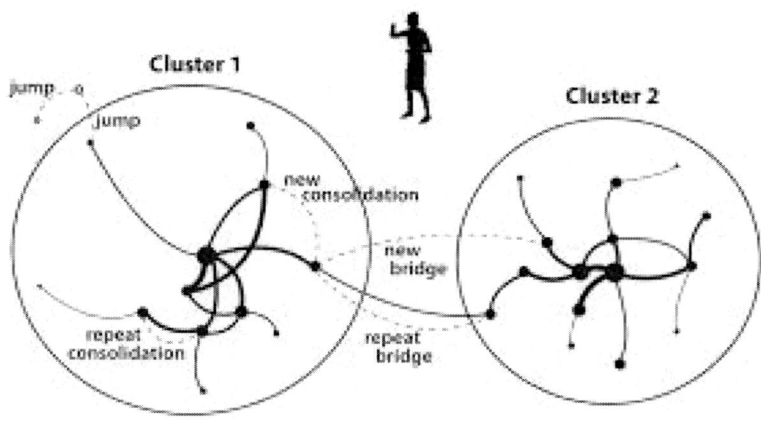

Figure 4.1. Scientific strategies on a network and between clusters within a network are depicted. Nodes represent chemicals, links represent chemical relations and the intensity represents the number of citations between papers. From [70] with permission.

and annotated by NLM with two or more chemical entities. They constructed an evolving SICN using the annotated articles, which over this 25 year period consisted of 181,078 nodes and 84,709,977 links. Figure 4.1 indicates the number of connections by means of citations, with a link becoming broader and darker as the number of citations increases. In this context they define a research strategy to consist of the following scientific relationships; *jump, new* or *repeat consolidation,* and *new* or *repeat bridge.* A *jump* beyond current knowledge, may consist of the design of a new pharmaceutical, or finding evidence of a new chemical receptor, etc. A researcher may also find *new* (not previously published) or *repeated* (previously published) relationships, which can be either *bridges* or furnishes *consolidations.* They go on to point out that this distinction is due to knowledge networks naturally partitioning into clusters, echoing STEM subfields. Entities within a given cluster, when joined, provide further *consolidation* of the cluster, intensifying knowledge in that subfield. A *bridge* between distinct knowledge clusters is created by links across cluster boundaries, thereby more tightly weaving what would otherwise be loosely connected subfields.

They quantified five measures of their research strategy and thereby tested their hypothesis that rare, risky strategies are more highly rewarded than conservative strategies, if and when such rewards occur. Figure 4.2 shows that rare and risky strategies, such as jumps and new consolidations, are indeed rewarded with more citations than conservative strategies, like repeating. The measure they developed uses information theory concepts. The self-information, or *surprisal*, is defined:

$$I(p_i) = -\log_2 p_i \,, \tag{4.1}$$

whose average value is the Wiener/Shannon information associated with observing outcome i of a discrete random variable. In the application of interest the discrete index refers to one of the five research strategies. Highly improbable outcomes are surprising, so the surprisal allots a large value to low probability outcomes. This measure quantifies how "surprising" such outcomes can be.

In the case of interest here, the outcomes are observations of a given research strategy and as expected the observations of rare (high-risk) strategies on the SICN are more surprising than those of more common (low-risk) strategies. Verifying the intuitive association of risk with strategy rarity was another positive outcome of their research. *They also determined that such rare, riskier strategies are less successful in attracting citations than are more conservative strategies in general.* However, when the rare, riskier strategies do attract more citations than do the conservative strategies, they can be spectacularly successful.

Foster et al. [71] reach the conclusion that STEM breakthroughs are so infrequent because the gamble is so great, with insufficient payoff on average in terms of citations garnered to justify the risk. This may explain, in part, why the greatest number of breakthroughs are made by the young. My interpretation of this young investigator effect (YIE) is that the young SI either does not know what is at stake, or knowing the risk still believe themselves to be up to the challenge and to be invulnerable. This is not unlike the young warrior in the military, whose actions win recognition and rewards, but does not guarantee their physical survival.

A significant limitation of the above analysis is the fact that the comparisons are only made among published papers, thereby excluding high-risk research that either produced no results, or results that could not survive the publication filtering process. In the context of S2TEM it might be possible to improve on

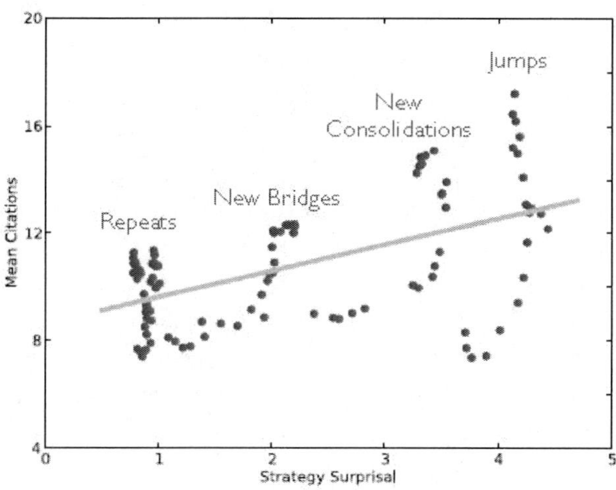

Figure 4.2. Strategies and citation impact are depicted. Low-frequency strategies are correlated with higher-mean citations (29% of variation explained). The surprisal of a strategy in a given year (horizontal axis) plotted against mean citations received by papers published in the same year containing at least one instance of that strategy. Much of the remaining variation is explained by adding subsequent years of observation to the model, as citations tend to increase with time (76% of variation is explained by the combined model; the surprisal coefficient remains significant). Adapted from [70] with permission.

these results by including in the analysis research proposals and government reports that were intended for broader publication, but were never successful in making it to the broader research stage. This could be a productive line of research for quantifying the true cost of high-risk research within a specific context.

4.2. What is a Breakthrough Idea?

In one sense a breakthrough idea in a STEM discipline is an atypical combination of accepted pieces of knowledge [72], in another sense it is the resolution of an empirical paradox (EP) [70] arising from data-based contradictions in accepted knowledge. It is the latter that we explore first. In broad outline every nascent field of STEM begins with elementary experiments, whose results are explained by simple models from which that discipline's basic concepts are extracted and developed further. As the STEM discipline matures the experiments become more sophisticated and those basic concepts are interwoven into ever more elaborate theories. This process of refinement and elaboration sometimes covers centuries, for example, theories on the nature of light spanned over two and one-half centuries from Newtons' *Opticks* to Feynman's *Lectures on Quantum Electrodynamics* (QED) and beyond.

Finally, a point is reached where a new experiment probes the previously hidden true nonsimplicity of a discipline's phenomena and in order to explain the new dataset a logical contradiction arises between two or more of the fundamental concepts of existing theory, resulting in an empirical paradox (EP). It is an EP because nature has presented us with something that logically cannot exist, based on existing knowledge, and yet it does. Consequently, the STEM discipline must reinvent itself to explain the new dataset and it does so by changing how investigators think about the phenomenon through a reinterpretation of the old datasets. Sometimes they extend existing theory, other times they jettison what exists (consider the now vanquished luminiferous aether thought to support the propagation of light through the vast regions of space between the stars) in order to invent totally new theory to resolve the paradox.

Let us begin with an example of an EP that, without exaggeration, totally changed how we understand the physical universe. Early in the twentieth century there existed a large body of empirical evidence establishing that light was a wave phenomenon. It was puzzling therefore that there also existed a significant amount of the then newly acquired data establishing that light was a particle phenomenon. The fact is that a particle is localized in space and a wave is extended in space, as depicted in the cartoon in Figure 4.3, and yet they both purport to explain the fundamental nature of light. These two data sets generated an EP, that being, the existence of a given phenomenon possessing contradictory, or mutually exclusive, theoretical properties.

Resolving this empirical paradox led to concept of wave-particle duality, the Copenhagen (quantum) interpretation of microscopic phenomena, with the result that the interpretation of what is observed is determined by the measurement that is made. Such measurement indicates that light is either a particle, or

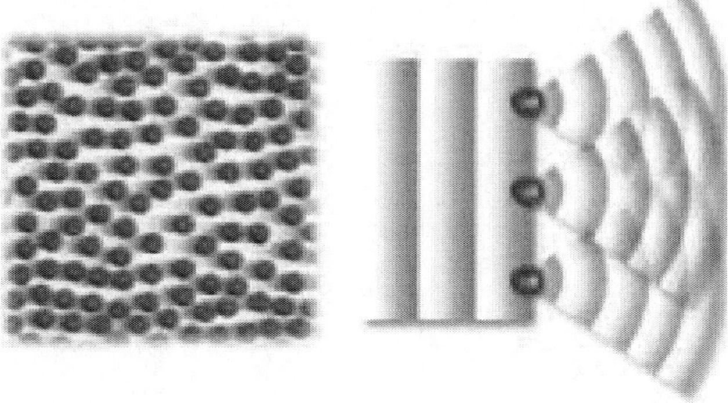

Figure 4.3. The nature of light in the first quarter of the 20th century had two competing theories: its nature is fundamentally particle-like as shown on the left, or wave-like as shown on the right. But according to classical physics both could not be true.

a wave, at any given time, but the data never indicate both simultaneously. The revolutionary interpretation of light using the quantum hypothesis was made in the 1905 paper of Einstein for which he received the Noble Prize in physics a decade latter and which led Niels Bohr and Werner Heisenberg in the 1920s to theorize that light is neither a particle nor a wave, but paradoxically it is both. Thus, this EP, interpreted under the rules of classical mechanics, entailed quantum mechanics for its resolution and a new way of knowing the universe was born. Note that this is not just new knowledge, but a new kind of knowledge, a kind of knowledge that would not be possible without the explicit quantum resolution of the wave-particle paradox.

Such dramatic changes in STEM disciplines are evident and are usually the result of a conscious decision made by a significant fraction of the most talented investigators within a subfield who work together to obtain the outcome. More subtle transformations of subfields often occur without such fanfare and may even elude discovery without concerted effort to find them. Uzzi et al. [72] determined that the highest-impact science results from a balanced mixture of atypical combinations of knowledge (EP resolution) and conventional domain-level thinking. Their purpose was to quantify this balance and determine how to achieve it using the network analysis of an appropriately defined SICN.

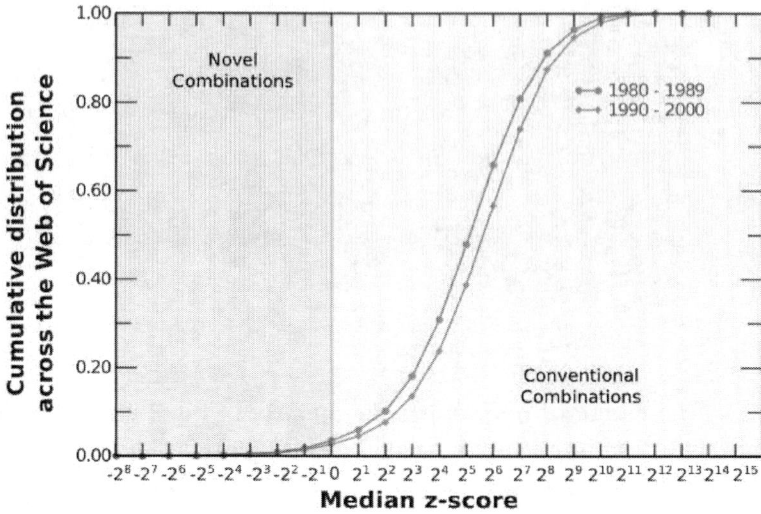

Figure 4.4. Novelty and conventionality in science. For a sample paper this figure displays the median z scores and shows that the vast majority of papers have a high propensity for conventionality; in the 1980s and 1990s, fewer than 4% of papers have a median z scores below 0 and more than 50% of papers have median z scores above 64. Adapted from [71] with permission.

They [72] generate a new kind of SICN by considering all pairwise combinations of references of each paper published within a given year in the Web of Science (WoS). The frequency of each co-citation pair across all papers in that year is then determined. The observed frequencies are compared with those expected using randomized SICNs of the same size, using a z-score for each journal pair. Positive z-scores indicate conventional or common pairings in which the data show pairings appear more often than would occur by chance. Negative z-scores indicate atypical or novel pairings where the data show pairings appear less often than would occur by chance. It is clear from the z-scores displayed in Figure 4.4 that the vast majority of papers have conventional combinations, whereas novel combinations are relatively rare. Perhaps even more rare than we would have predicted, given our bias for how original we see STEM research in our mind's eye.

A surprising observation made by Uzzi et al. was that team-authored papers are more likely to contain atypical combinations than do either single-authored or pair-authored papers. They conclude that novelty and conventionality are not opposing but complimentary factors in science productivity [72]:

> At root, our work suggests that creativity in science appears to be a nearly universal phenomenon of two extremes. At one extreme is conventionality and at the other is novelty. Curiously, notable advances in science appear most closely linked not with efforts along one boundary or the other but with efforts that reach toward both frontiers.

The term science used by Uzzi et al. did not make the distinction made by using the term STEM and therefore the WoS data contained citation data from science, technology, engineering and mathematics. There is every expectation that similar results will be found using a rigorous STEM dataset, but apparently such statistical analysis remains to be done.

Quality versus Popularity

It is well-known that when a Noble Laureate quotes a paper other STEM investigators pay attention, more so than when a typical researcher cites the same paper. This *great scientist effect* (GSE) has not, so far, been included in our discussion of SICNs, but it can be a strong influence in the development of patterns, both small and transient, as well as, large and persistent. Often times the difference between a significant and a merely popular paper is not the number of citations the paper receives, but who is doing the citing [73]. In point of fact, one paper may have fewer citations than another and yet the less cited paper can be of higher STEM quality (significance) as anticipated by the GSE.

Zhou et al. [73] point out that the previous studies concentrated on the rankings of investigators as well as their papers, while at the same time neglecting their mutual dependence. These authors elected to incorporate this interdependence by means of a SI-paper bipartite network, consisting of the SICN and an SI-paper network (SIPN). The method is parameter-free and simultaneously yields rank lists of authors and papers.

The SICN-SIPN bipartite network (SSBN) incorporates the reputations of SIs, publications, as well as their interactions. The links in and out of each node on the SSBN enables the prestige of an author to be included in the strength of the link. They [73] argue that the diffusion from papers to authors on the SSBN is a conservative process, however that from author to paper is non-conservative. The non-conservation follows from the fact that two papers with the same number of citations are not necessarily weighted equally because the more prestigious researchers carry more weight in their cites. A SI's citation on the SSBN is the total number of citations received by that SI from other papers. A paper's citation index is the number of papers that cite it.

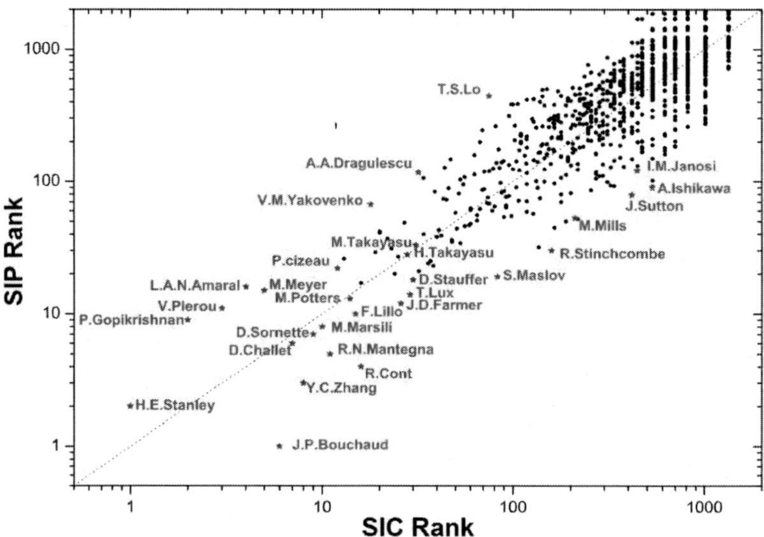

Figure 4.5. Scatter plot of SIP rank versus SIC rank for authors. If the two methods provide the same ranking all the points would fall on the diagonal. The outliers indicate the significant difference between SIP rank and SIC rank. Some typical examples are labeled in red. The Kendall correlation coefficient is 0.784. Adapted from [72] with permission.

Figure 4.5 displays the scatter plot of SI (authored) paper (SIP) rank versus SI citation (SIC) rank for SIs in the field of econophysics. If the two ways of counting produced the same rank then all the data points would lie along the diagonal, thereby indicating that quality and popularity would be the same, which is clearly not the case. It does however, show that the SIP rank ordering is roughly proportional to the SIC rank ordering, with a correlation between the two ways of counting of 0.784. This high level of correlation indicates that although not equivalent, quality and popularity are not independent of one another.

It is clear from the figure that Professor Gene Stanley, who along with Professor Mantegna founded the discipline of *Econophysics*, is ranked #2 in SIP rank and #1 in SIC rank. The SIs with scores below the diagonal have a higher SIP rank (lower number) and have a lower rank (higher number) as measured by the SIC rank score.

Journal Rankings

Most scientists I know would give a sizable fraction of a month's salary, above the cost of page changes, to have a paper published in *Nature*. The reason for that is, at least in part, the belief that their research would take on the veneer of quality associated with that venerable journal, which has an impact factor (IF) of 43 (in 2020). The IF is typically defined as the average number of citations per year a paper published in that journal receives over a two year period. However, as we pointed out with regard to the $h-$index, because this is based solely on the number of citations it is more a measure of popularity than it is a measure of quality. But scientists, being the fragile human creatures they are, would settle for the STEM popularity of being published in *Nature*, at least in the short term, as opposed to waiting for the quality of their research to be recognized.

Zhou et al. [73] conclude, using the above arguments on popularity, that the IF is not sufficient to measure the science quality or prestige of a journal. They elect to compare the IF to a separate and distinct measure, the *Article Influence Score* (AIS). Bergstrom [74] explains that the AIS takes into account that fact that a single citation from one high quality journal may be of higher value than multiple citations from lower quality journals, thereby extending to journals the recognition made to the citation prestige of individual researchers made in critiquing the $h-$factor earlier. Thus, just as citations from different SIs carry different weights, so too do the citations from different journals. Consequently, an influential journal is one that is cited by other influential journals. The circular nature of this argument is overcome by iteratively calculating the importance of each journal with a simple mathematical algorithm Bergstrom named *Eigenfactor*. A 5-year average IF and AIS measures of influence are compared in Figure

4.6 for 18 of the better known STEM journals, including *Science* and *Nature*. Each journal is ranked according to their AIS in descending order. With few exceptions the two measures of each journal are positively correlated. Other than these few anomalies the two measures could be used interchangeably. So unless one is the editor of one of these anomalous journals and interested is increasing the journal's AIS, one might argue that the IF is a sufficient measure of journal quality for most purposes.

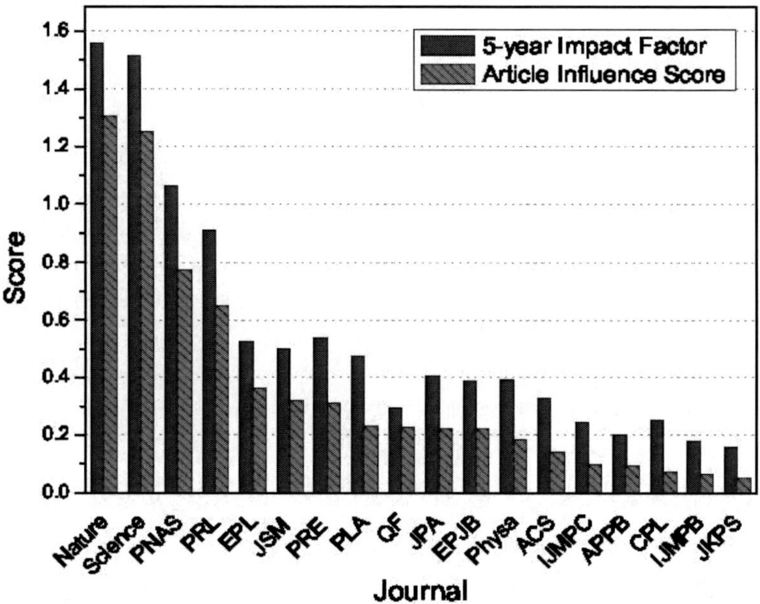

Figure 4.6. The 5-year IFs and AISs of eighteen journals are in the dataset for the plot. For clearer presentation, the scores are presented in the form $log_{10}(s+1)$, where s is the real value of 5-year IF or AIS. From [73] with permssion.

Chapter 5

Universality

Many nonsimple dynamic networks have well-defined scaling structures, and the scaling exponents are found to be invariant with respect to small changes in the network parameters, thereby allowing scaling phenomena to be classified in terms of scaling exponents. As explicitly pointed out by Meakin [75] networks with the same scaling exponents belong to the same universality class, meaning that their underlying dynamic models are not the same and yet they behave identically in the vicinity of a critical point. It is the renormalization group flow away from the critical fixed point that controls critical behavior not the detailed structure of the dynamic equations [76]. Moreover, such discussions of universality as used here have historically been associated with the physical sciences through phase transitions with dynamics in the vicinity of a critical point, for example, fluids and magnets changing their macroscopic properties as the temperature is lowered through its critical value.

More recently criticality has been found in the context of modeling nonsimple phenomena using Network Science in which the network dynamics are members of the same universality class, such as the Ising Model [75, 76] and the Decision making Model (DMM) [28]. Nonsimple biological systems have been found to be poised at criticality [77] and similar universal behavior has been uncovered in social phenomena [28, 49] as well. This emergent behavior is a kind of macroscopic dynamic stability against changes in the microscopic dynamics. In this chapter we exploit the general nature of the interdependence among the properties of universality, scaling, and nonsimplicity to develop a deeper understanding of the nonsimple networks that are formed in carrying out STEM research. We begin with a principle or law associated with a well-defined observable and demonstrate how such an empirical regularity follows from theoretical arguments. One such law is "form ever follows function" and in its most poetic articulation is given at the close of the 19th century by its author [78]:

> Whether it be the sweeping eagle in his flight, or the open apple-blossom, the toiling work-horse, the blithe swan, the branching oak, the winding stream at its base, the drifting clouds, over all the coursing sun, form ever follows function, and this is the law. Where function does not change, form does not change. The granite rocks, the ever-brooding hills, remain for ages; the lightning lives, comes into shape, and dies, in a twinkling.
>
> It is the pervading law of all things organic and inorganic, of all things physical and metaphysical, of all things human and all things superhuman, of all true manifestations of the head, of the heart, of the soul, that the life is recognizable in its expression, that form ever follows function. This is the law.

This law indeed generalizes to the situation where form and function change over time, but the relation between them remains the same. A well-known empirical connection between a phenomenon's form denoted by its size and its functionality is determined by an empirical allometry relation (AR). West [79] has shown that the basis for the ubiquitous nature of empirical ARs, found in virtually every discipline from anatomy to zoology, is the nonsimplicity of the underlying dynamic network, thereby resulting in a PDF that scales.

van Raan [80] used an analogy with the urban AR studies of Bettencourt [81] to empirically establish the existence of an AR for the 500 largest universities in the world for which the number of citations is the functionality and the number of publications is the size. He notes the basis of the power-law relation between citations and publications to be a consequence of the nonsimplicity of network dynamics. In this chapter, we take his observations a step further using the properties of renormalization group theory and the statistical properties of bibleometric indicators which van Raan [57] determined to have co-citation clusters described by fractal statistics, as indicated in Figure 3.5. In general the formation of allometry-conjugate pairs of variables have been shown [79] to be the result of how the nonsimplicity of a network increases monotonically with size and the functionality of that network responds monotonically to increasing nonsimplicity.

5.1. C-AR

What general conclusions can we draw from the myriad of measures considered so far? One such conclusion regards the universality of the citation allometry relation (C-AR) and in this chapter we present a mathematical argument in support of the C-AR, which Molinari and Molinari [66] (MM1) introduced heuristically as the master-curve. The universality is associated with the dynamic properties of the interactions among the elementary elements constituting the nonsimple dynamic network. In physical systems undergoing phase transitions the universality of the dynamics is captured by the scaling of the emergent behavior of the macroscopic dynamics. For example, the change of phase from liquid to solid as one lowers the value of the control parameter (temperature) thereby inducing a transition from short-range to long-range interactions.

The universality appears in the functional dependence of the scaling index on the control parameter. The relatively short-range correlation of the fluid fluctuations is extended as the critical value of the control parameter is approached. In the physical domain the autocorrelation function is an exponential in the subcritical domain but transitions to an IPL near the critical point where the IPL index becomes singular. The critical behavior of nonsimple dynamic networks is universal, so that if the STEM research networks being discussed are in fact nonsimple in the same way as in a physical network we expect their statistical properties to be universal. But are they?

Herein we use the nonsimplicity of social interactions in research activities to establish a C-AR between a measure of the functionality of the investigative research process and the size of the extended network involved in the research. As the size of a network increases more opportunity for variability becomes available, thereby influencing the functionality of the network. The statistical variability of a variate's dataset is one measure of nonsimplicity [82], but it is not the only measure. In the present context a research network could be formed through the citations of papers within and across STEM disciplines, with collaborations over time among the investigators publishing papers, or any of several ways discussed to characterize the elements of a STEM research network [49, 50].

The basis for the ubiquitous nature of allometry relations (ARs) has been shown [79] to be the underlying nonsimplicity of the networks. This nonsimplicity results in an average functionality of the network being related to the average network size through an AR. It has been demonstrated that an AR is in general a consequence of the scaling behavior of an underlying PDF. It is not surprising then that an activity as nonsimple as that of interconnecting the STEM research developed in a thematic area through a citation network over time should asymptotically be captured by a C-AR. West and West [83] connected the scaling of the PDF, characterizing the statistics, to the relation between the average functionality $\langle Y \rangle$ and average size $\langle X \rangle$ of a nonsimple network through the monotonic AR:

$$\langle Y \rangle = a \langle X \rangle^{b}, \tag{5.1}$$

where a and b are empirical constants in the simplest case. Empirical ARs have been established for such allometry conjugate pairs (Y, X) as: in life history [84, 85] (metabolic rate, body mass), (brain mass, body mass), (biological time, body mass), (white matter mass, gray matter mass), (wing span, body mass); in urban growth [81] (innovation, urban population),(total wages, urban population); in sociology [86] (mortality rate, population density); in hydrology [87, 88] (basin length, basin area); in STEM research [66] ($h-$index, total number of papers); as well as in a wide variety of other disciplines [79]. The formation of such conjugate pairs has been shown to be the result of how the nonsimplicity of a network increases monotonically with size and the functionality of that network responds monotonically to the increasing nonsimplicity.

Scaling

As the size of a network increases more opportunity for variability becomes available, thereby influencing the function of the network. As explained earlier, a research network can be formed through the citations of papers within and across STEM disciplines, the collaborations over time among the investigators publishing papers, or any of a number of ways to characterize the behavior of a research network. In order for a network to retain its functionality as its size continues to increase, macroscopic dynamic modes must emerge to replace those that no longer support the network's evolving function driven by the increase in nonsimplicity. The network size and functionality increase and decrease together as determined by their separate connection to the changes in nonsimplicity, but not in direct proportion to one another. The detailed underlying dynamics of the network is not important due to the criticality of the interactions, rather it is the emergent properties that determine the macroscopic control using nonsimplicity as the intermediary.

Another way to view the relation between nonsimplicity and size is by relating the functionality of the collective phenomenon of interest to the network's nonsimplicity. West [79] was among the first to identify that the more sophisticated the functionality, the greater the nonsimplicity necessary for the network to carry out that function. Consequently, we interpret the many empirical relations, between functionality and size, as being the result of the implicit relation between size and nonsimplicity, with the latter being manifest through the network's functionality. This subtle, yet ubiquitous, driving of nonsimplicity by size and functionality by nonsimplicity, was hypothesized to form the foundation of the science of allometry [79], which ultimately relates functionality to a non-integer power of size in the form of an AR. The generality of this argument suggests that it is not restricted to any particular STEM network and in fact we find such ARs in ecosystems [89, 90], urban development [81], accidents [91], locomotion [92], and the evolution of technology [10] to name just a few more.

SICNs

As stated elsewhere [79] the increasing size of a network provides increasing nonsimplicity that is necessary to maintain network stability. Mandelbrot [88, 93] identified a number of AR models masquerading under a variety of heuristic *laws*. He argued that these laws are a consequence of nonsimple phenomena not having characteristic scales, that is, the phenomenon associated with the law manifest scaling and are fractal. Scaling is a ubiquitous property of large nonsimple networks indicating that the observables with which such networks are characterized fluctuate over many scales of the independent variables in a coordinated manner. An observable $Z(t)$ is said to scale if it satisfies the homogeneous scaling relation:

$$Z(\lambda t) = \lambda^{\beta} Z(t), \tag{5.2}$$

where the scaling index is β.

A relatively simple index that characterizes the quality of a network of STEM research publications, whether the 'research network' is that of a university, of a laboratory, or of a single investigator, is the Hirsch h-index [59], which we introduced in Chapter 3. The increase in the number of citations drives the nonsimplicity of the research network. Consequently, in the present case the observable Z is interpreted as the h-index and t denotes the number of papers n with h or more citations and the scaling has to do with the PDF and not with the dynamic citation variable itself.

Consider the PDF $P(h,n)$, which yields the probability that the h-index is in the interval $(h, h+dh)$ for a given number of papers n. When a complex network can be shown to have critical dynamics and consequently if the citation network is of this form the citation PDF satisfies a fractional diffusion equation (FDE) whose discussion here would take us too far from our main purpose, so we refer the reader to [34] for details. The solution to this FDE can be obtained using renormalization group theory to obtain a PDF of the form [79]:

$$P(h,n) = \frac{1}{n^{\beta}} F\left(\frac{h}{n^{\beta}}\right), \tag{5.3}$$

and $F(\cdot)$ is an unknown PDF of the scaled variable. If this were a simple diffusive process the unknown function would be a Gaussian distribution with $\beta = 1/2$. However, for networks with $\beta \neq 1/2$ could also have a Gaussian PDF, in which case the process is one of fractional Brownian motion (FBM). In the more general case the statistics are not Gaussian.

We can write the average h-index by inserting Eq.(5.3) into the integral for the average:

$$\langle h;n\rangle = \int hP(h,n)dh = h_m n^{\beta}, \tag{5.4}$$

where h_m is the impact factor and the average $\langle h;n\rangle$ is associated with $\overline{h}$, both of which are defined in MM1 and discussed in Chapter 3. Eq.(5.4) is fit to data in Figure 5.1 where $\beta \approx 0.4$ for asymptotic $n = N$ and the impact factor h_m is the intercept of the solid line segment with the vertical axis.

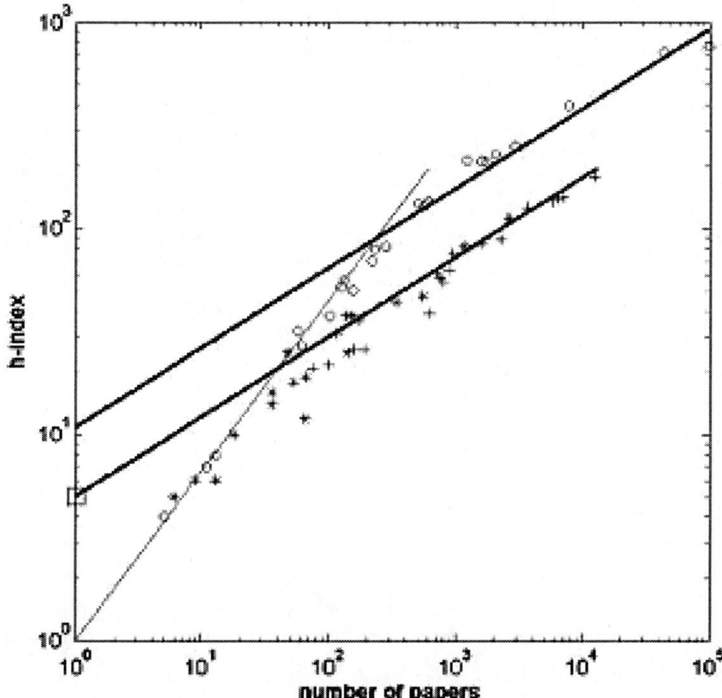

Figure 5.1. Master curves of scientific journals: *Science* (dots), *Acta Materialia* (crosses), the *Journal of the Mechanics and Physics of Solids* (asterisks). These master curves show the evolution of the h-index in terms of number of papers. Each data point in this log–log diagram corresponds to a given country. The thin and bold straight lines are a fit to the data and distinguish two different growth regimes, $\beta \approx 0.8$ for $n < 200$ and $\beta \approx 0.4$ for $n > 200$. The square represents the intersection of the lowest bold line with the vertical axis and characterizes h_m for *Acta Materialia* (or for the *Journal of the Mechanics and Physics of Solids*). From MM1 with permission.

The homogeneous scaling relation given in Eq.(5.2) takes the form for the average h-index given by:

$$\langle h; \lambda n \rangle = \lambda^{\beta} \langle h; n \rangle \,, \tag{5.5}$$

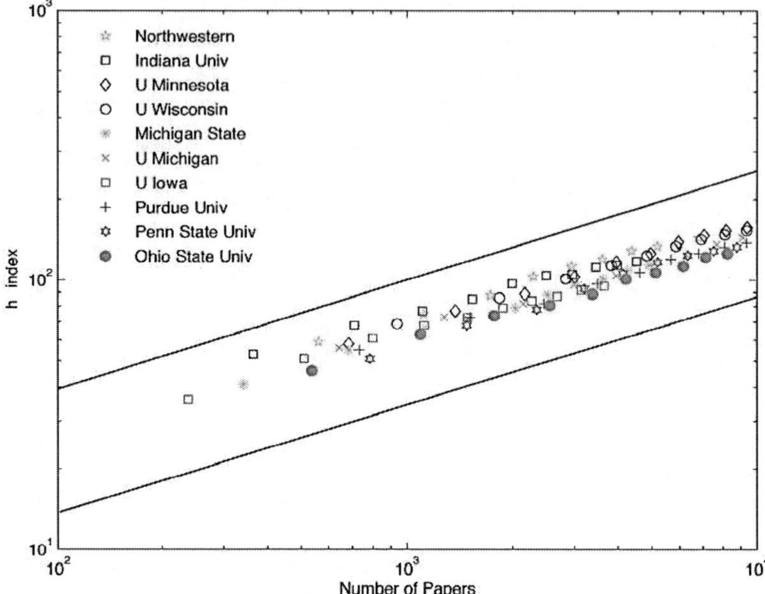

Figure 5.2. Master curves defined by Eq.(5.4) depicts the $h-$index for a given number of non-medical publication for ten of the top U.S. research universities. A remarkable aspect of these measures is the narrow dispersion in the impact factor for these universities. The parallel lines enclosing the data are intended to aid the eye in determing the slopes to all be in the vicinity of $\beta \approx 0.36$. This figure is from [93] with permission.

from which by setting $\lambda = 1/n$ and comparing the resulting equation with Eq.(5.4) yields the impact factor:

$$h_m = \langle h; 1 \rangle , \tag{5.6}$$

which is the intercept with the vertical axis given by the square in Figure 5.1, as identified in MM1. Here the impact factor is alternatively determined by an average of the scaled variable $q = h/n^{\beta}$ over the unknown distribution:

$$h_m \equiv \int qF(q)dq. \tag{5.7}$$

MM1 referred to Eq.(5.4) as the master-curve which they interpreted as the signature of the impact of a given large set of research papers, for $n > 200$. Note that the master-curve has the C-AR form given by Eq.(5.1) and that the theoretical form of h_m given by Eq.(5.7) is independent of n. Moreover, h_m is interpreted to be the level of influence the indicated journal in each country has on the scientific community.

Kinney [94] published a study while MM1 was still in press supporting the MM1 findings that there exists a universal growth rate for large numbers of papers enabling significant comparisons of the research quality between institutions. Figure 5.2 is one of many such comparisons made by Kinney and depicts the $h-$index for a given number of non-medical publications for ten of the top U.S. universities. Note how tightly the empirical data cluster around the master curve given by the two purple curves to aid the eye. This figure is typical of the results found.

Master Curve

The sequel to MM1 answers a number of questions related to the new measure not answered in MM1. In particular, they [95] (MM2) analyze the relations between the underlying rank/citation-frequency law and the h-index. Here we exploit the scaling assumption leading to the present form of the master-curve in terms of the C-AR to provide an overall fit to the rank ordering of the citation-frequency law. Note that a set of N_T papers can be ordered according to rank, such that the jth paper has $n_c(j)$ citations, which are rank ordered according to $n_c(1) > n_c(2) > ... > n_c(N_T)$. MM2 assume that a continuous PDF $(x, g(x))$ can to a good approximation replace the discrete pair $(j, n_c(j))$ and that in keeping with Zipf's law they assume the citation frequency to have the IPL form:

$$g(x) = \frac{c}{x^{\alpha}}, \tag{5.8}$$

where c is determined by normalization and x is the rank order of the paper, often denoted as r in the mathematics literature. They use the approximate IPL form of the rank-order PDF given by Eq.(5.8) along with the C-AR to determine

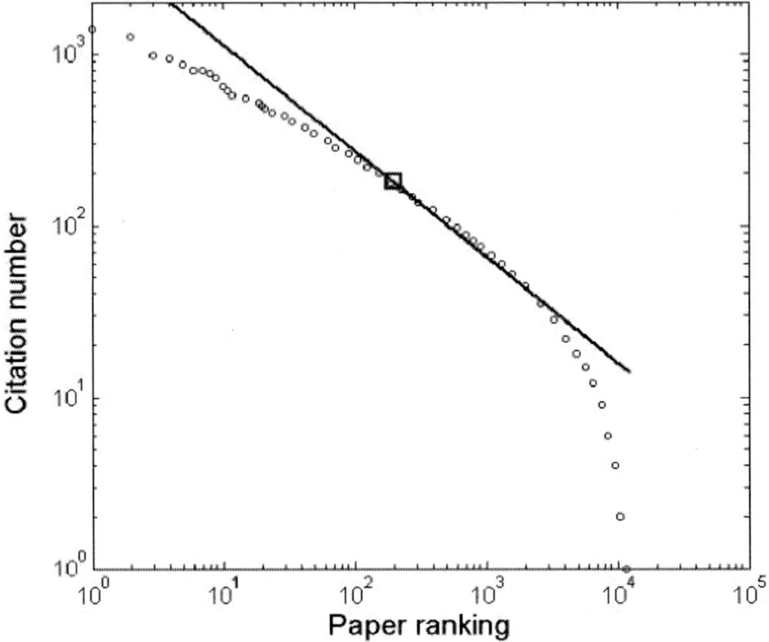

Figure 5.3. The square represents the point where the h-index is reached at the intersection of the bisector with the graph. The heavy line is a best fit straight line to the data in the region $10^2 \leq x \leq 10^3$ with a slope of $\alpha = 0.62$. This line represents reasonably well the citation law in a range of citations including the h-index. From [94] MM2 with permission.

the allometry index β in terms of the IPL index α is:

$$\beta = \frac{\alpha}{1+\alpha}. \tag{5.9}$$

The fit of the IPL to the data in the region $10^2 \leq x \leq 10^3$ depicted in Figure 5.3 from MM2 yields $\alpha = 0.62$ with a resulting exponent for the C-AR of $\beta \approx 0.38$ from Eq.(5.9).

The fact that $\beta < 1$ implies an economy of scale. This economy of scale associated with the h-index of a citation network means that the more papers published the greater the cost savings per citation. This would mean with a suitable modification of interpretation to take into account the non-monetary nature of the network, that the efficiency of knowledge transmission with this network is enhanced with the increasing number of papers published.

5.2. Mittag-Leffler Probability

The rank-order distribution is typically given in the literature as an IPL in the manner of Zipf's law. However, the IPL form of the rank ordering is almost always an approximation to a more exact solution to a fractional rate equation such as the one developed below in terms of the Mittag-Leffler function (MLF). The manner in which the MLF is derived is based on an application of the fractional calculus to the dynamics on the STEM research network. The details can be found in West [79] but what the fractional calculus entails about the underlying network dynamics in the present context is given below.

The simple random behavior of an individual, when isolated, whether the individual is a person or an organization, is replaced with behavior that might serve a more adaptive role when introduced into a large network of similar individuals. Turalska and West [96] conjectured that the adaptive behavior adopted by an individual due to its interaction within a dynamic network of similar individuals is generic, given that the dynamics of the nonsimple network belongs to the Ising universality class [28]. Members of this universality class share critical temporal behavior that drives a subordination process. It is the renewal property of the event statistics, which, through the subordination process, gives rise to the linear fractional rate equation (FRE) for a typical individual's society-induced dynamics. The solution to the FRE manifests the subsequent robust statistical behavior of the individual.

We assume here that the citation network due to its nonsimplicity, as manifest in its critical dynamics, has a FRE in terms of the rank-ordering parameter x for the citation frequency probability $\Psi(x)$:

$$\partial_x^{\alpha}[\Psi(x)] = -\lambda\Psi(x). \tag{5.1}$$

Here $\partial_x^{\alpha}[\cdot]$ is the Caputo fractional derivative, with $0 < \alpha \leq 1$ and $\lambda > 0$ [97].

The solution to this FRE is given by the Mittag-Leffler function (MLF) [79]:

$$\Psi(x) = c' E_{\alpha}(-\lambda x^{\alpha}) = c' \sum_{j=0}^{\infty} \frac{(-\lambda x^{\alpha})^{j}}{\Gamma(j\alpha + 1)}, \tag{5.2}$$

and c' is a constant. In the present discussion of the rank-order of the citation number there are x publications with probability $\Psi(x)$ that the relative frequency of values is greater than or equal to the number of citations. The FRE is a generalization of a Poission process and reduces to it when $\alpha = 1$ [98, 79], in which case the series expansion of the MLF given by Eq.(5.2) sums to an exponential.

An IPL rank-order distribution is fit to the dataset shown in MM2, see Figure 5.3. A fit to the MM1 dataset using a properly normalized MLF is depicted in Figure 5.4. It is evident the MLF fits the dataset over *three and one-half decades* with a quality of fit of $R^2 = 0.996$, whereas MM2 fit the same rankings over a *single decade* using the approximate IPL.

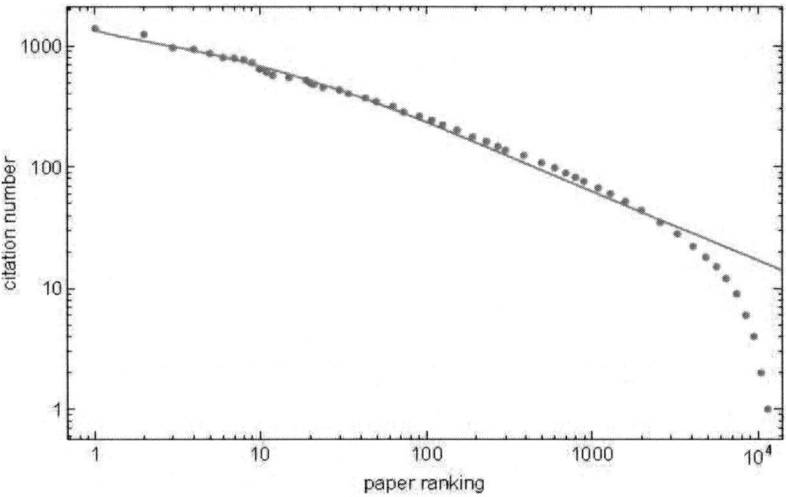

Figure 5.4. A log-log plot of the properly normalized rank-order citation MM2 dataset. The solid curve is a MLF given by Eq.(5.11) with parameter values fit to the dataset: $\alpha = 0.564$ and $\lambda = 0.225$,with a quality of fit $R^2 = 0.996$. The dataset was generously provided by the authors of MM1.

Note that the asymptotic form of the MLF is given by [98, 79]:

$$\lim_{x\to\infty} \Psi(x) = \frac{c}{x^{\alpha}}, \tag{5.3}$$

where $c = c'/\Gamma(1-\alpha)$ in formal agreement with the approximate IPL fit used in MM2. Here the MLF is the exact rank-order distribution to which Zipf's law is an approximation. We see that the IPL index of this asymptotic form of the MLF enables us to use this parameter to estimate the universal C-AR index to be $\beta = 0.36$ from the overall fit of the MLF to the dataset which yielded $\alpha = 0.564$. It is clear that the value of β calculated using the full MM1 dataset remains close to the value 0.4 as estimated in MM2 using just the IPLs.

5.3. Hyperbolic PDF

A related method for determining the C-AR index uses the Pareto PDF and not Zipf's rank-ordering probability. In this section we construct and solve the steady state Fokker-Planck equation (FPE) for the PDF of interest. The motivation for this approach is the insight gained through Shockley's argument of the multiplicities nature of research publication complexity. Given an observable related to the publication of, or citation to, a research paper, that observable would be stochastic given its dependence on the nonsimplicity of STEM research networks and the variability of the skills required to publish a paper. Note that in Chapter 3 we identified the PDF of the aggregated number of citations associated with an institution is hyperbolic.

Multiplicative Fluctuations

To model the dynamics of a research observable $Z(t)$ including its statistical fluctuations we assume it satisfies the multiplicative Langevin equation:

$$\frac{dZ(t)}{dt} = G(Z) + g(Z)\xi(t), \tag{5.1}$$

where $\xi(t)$ is a Wiener random process of strength D and correlation function:

$$\langle \xi(t)\xi(t') \rangle = 2D\delta(t-t'). \tag{5.2}$$

This Langevin equation describes a positive definite process whose regular dynamics are determined by $G(Z)$ and the coefficient of the random fluctuations is the state-dependent function $g(Z)$, where $Z(t)$ is an as yet unspecified function of the number of citations.

The details leading to the corresponding FPE for the PDF are provided in [79]:

$$\frac{\partial P(z,t)}{\partial t} = \frac{\partial}{\partial z}\left[-G(z) + Dg(z)\frac{\partial}{\partial z}g(z)\right]P(z,t). \tag{5.3}$$

We are not aware of any closed-form solution for this FPE. However, it can be solved asymptotically to obtain the steady-state solution:

$$P_{ss}(z) = \lim_{t\to\infty} P(z,t).$$

Imposing the zero flux condition on the steady-state FPE:

$$\left[-G(z) + Dg(z)\frac{\partial}{\partial z}g(z)\right]P_{ss}(z) = 0, \tag{5.4}$$

allows us to obtain the normalized steady-state PDF:

$$P_{ss}(z) = \frac{C}{g(z)}\exp\left[\frac{1}{D}\int^{z}\frac{G(x)dx}{g(x)^2}\right], \tag{5.5}$$

where C is the normalization constant.

Applying the conditions for a stable SICN to the steady-state solution given by Eq.(5.5) we select the phase-space variable to be $z = N + n$, where $n \geq 0$ is the number of citations and N is a positive real number. We select the deterministic function to be:

$$G(z) = -\lambda g(z), \tag{5.6}$$

and the multiplicative function to be:

$$g(z) = N + n. \tag{5.7}$$

These functions together specify a Langevin process where the linear growth rate is stochastic, and is attenuated by a deterministic linear dissipation. Inserting these functions into Eq.(5.5) and normalizing yields the hyperbolic PDF

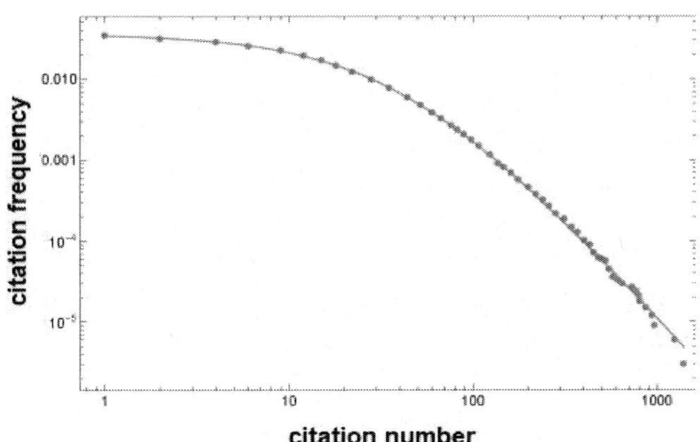

Figure 5.5. Plot of the citation-frequency function or PDF $p(n)$ yielding the probability that the number of citations is in the interval $(n, n+dn)$ on a log–log scale. The solid curve is the hyperbolic PDF given by Eq.(5.20) with parameter values fit to the data: $a = 1.53$, $N = 42.67$ and a quality of fit given by $R^2 = 0.9997$. Data provided by MM2.

$p(n) = P_{ss}(n)$:

$$p(n) = \frac{aN^a}{(N+n)^{1+a}}, \tag{5.8}$$

and the IPL index is $a = \lambda/D > 0$.

The hyperbolic PDF given by Eq.(5.8) gives a two parameter theoretical fit to the MM2 dataset. Here again we use the full range of the dataset to graph the number of papers having a given citation frequency $(p(n), n)$ to obtain the PDF depicted in Figure 5.5. The curve fitted to the dataset is the hyperbolic PDF given by Eq.(5.8), which clearly has the asymptotic form of a Pareto IPL. The fit of the hyperbolic PDF to the MM2 dataset is truly remarkable. The parameter values fit to the dataset are $a = 1.53$ and $N = 42.67$, with a quality of fit to the data given by $R^2 = 0.9997$.

In the asymptotic domains where both the rank-order probability (Zipf) and the citation frequency PDF (Pareto) distributions are IPL we obtain $\alpha = 1/a$. Consequently, we obtain an estimate of the rank-order IPL index $\alpha = 0.65$ with a corresponding estimate of the C-AR exponent of $\beta = 0.39$, essentially reproducing the results of MM2. Note that the result given here includes the influence of the entire dataset, whereas the results obtained in MM2 rely on a restricted domain of the IPL approximation of the dataset.

The relation between the IPL indices in the two cases is a consequence of interchanging the axes [99]. Adamic [99] points out that the phrase "The rth largest city has n inhabitants" is the same as saying, "r cities have n or more inhabitants", which is equivalent to the definition of the Pareto PDF, except the vertical and horizontal axes are flipped. He goes on to prove that this is all that is necessary to obtain the parameter equality $a = 1/\alpha$.

Chapter 6

The Study

In this chapter we review the results of a STEM research study[1] that enabled the formulation of a preliminary answer to the two question posed in the first chapter, which we repeat at the appropriate places in the present discussion. The study [100] focuses on how a set of measures of STEM research quality, such as those discussed, hinted at, and suggested in Chapters 2 through 5, can be quantitatively determined. The study was in response to the ARO RFP designed to determine the relative research value of a variety of STEM research projects generated using three distinctly different ARO STEM research funding strategies. The study analyzed datasets selected from projects funded by ARO through SI, MURI, and UARC award mechanisms, as well as, by related datasets in the STEM literature. The datasets analyzed contained projects whose outcomes spanned the 10-year period starting in 2005 and ending in 2014. The objective of the study was to identify and characterize any systematic, relative difference in the value of the research results generated by principle investigators (PIs) based on the prestige of the PI's institution, as well as on the funding mechanism used.

The three dimensions were identified to characterize how the basic STEM research supported by ARO generates value for society in general and the U.S.

[1]The next three subsections on indicators draw heavily from the RTI report [100] in order to establish the measures discussed in the numerical results section. But it must be emphasized that the conclusions drawn from the numerical results along with the interpretations presented are strictly those of the author and do not necessarily represent the views of RTI nor ARO.

Army in particular and are *Scientific Significance* (*SS*), *Mission Relevance* (*MR*), and *Breakthrough Potential* (*BP*). They were framed as if the three dimensions of STEM research were essentially orthogonal and uncorrelated, but their interdependence was recognized, so that trying to collapse them into a single index would have been misleading. The value of a STEM research result is measured by devising indicators that locate its position in the three-dimensional space of STEM research values. The analytical approach to assessing research results along these three axes in isolation, and then investigating correlations through a sundry of analytical outputs is described in detail elsewhere [100].

In this chapter we briefly review the partitioning of each of the three dimensions of STEM research value into six indicator functions that are accessible using the same dataset. In Section 6.1. the *SS*-dimension of STEM research value is discussed in terms of the six measures: Immediate Visibility, Longer-term Visibility, Field Citation Ratio, Degree Centrality, Journal Visibility and Continuity of Effort. Section 6.2. addresses the *MR*-dimension in terms of the indicators: Concurrent DoD Investment, Follow-on DoD Investment, Social Media Mentions, ARL Co-authorship, Army Citations, and DoD Citations. Section 6.3. partitions the *BP*-dimension into: Patient Citation, Government Interest in Patents, Interdisciplinarity, Corporate Authorship, Corporate Investment, and Corporate Citations.

Section 6.4. sketches a number of the results from the study using the SI and MURI funding strategies relative to ISN within an 18 dimensional space. The final section interprets the results presented in the context of the two questions concerning funding strategies and makes contact with the techniques discussed in earlier chapters.

6.1. Scientific Significance as a Source of Value

In terms of fundamental STEM research, findings are judged based on the soundness of the research that produced them, along with their implications for future research. A discovery has little value if it is based on underlying STEM findings that cannot be confirmed by independent experiments and/or theory. This is a criticism often leveled against pieces of STEM research in the social sciences. Its value may also be minimized if other STEM researchers choose, for whatever reason, to ignore it. Therefore, the value of a STEM research result depends in part on whether the result is acknowledged and accepted as significant by the appropriate STEM community.

This is the basis for methods of STEM research assessment involving citation analysis. A citation to an article by another research article has no inherent value, but it provides a signal that other members of the STEM community might well appreciate the significance of the original article, and support its influence on downstream research.

Scientific Significance (*SS*) Indicators: *SS* is defined as being research that contributes substantially to knowledge in a particular field of study, so the term scientific in *SS* actually means any of the four areas in the STEM acronym, and *SS* actually means STEM significance. The quantitative measures developed are based on current best-practice techniques in bibleometrics, where patterns of how the STEM articles published based on the ARO-funded research are cited by other researchers. The indicators follow established guidelines in bibleometric assessment, such as the *Leiden Manifesto* [101]. The constructed measures of *SS* provide proxy measures indicating how research results have influenced subsequence research. The research value framework is meant to reflect aspects of *SS* other than impact, and are not directly measuring STEM merit or the STEM research quality of ARO-funded research, but is used to establish a relative measure between funding strategies, as will subsequently be made clear.

Bibleometric measures have recently been applied as indicators to assess STEM researchers, research efforts as well as research organizations. The view adopted in this study was that bibleometrics and citation indicators do not provide absolute measures of the quality of STEM research efforts, nor of researchers and research organizations. They measure three components of the *SS* dimension: *Visibility*: to what extent was this PI's research noticed by peer STEM investigators? *Influence*: how did the PI's research shape the activities of the STEM community, and was it influencing key individuals? *Persistence*: did the visibility and influence of the STEM research continue and grow over time, or was its visibility short-lived?

Visibility Indicators

Citation analysis is the most obvious indicator of the popularity of a publication, since it is the most visible indication of the response of the STEM community to a given research result. As the citation number builds over time the evident visibility of the research seems to provide an irrefutable measure of the research value of the work to others working in related areas. Using traditional bibleometrics, it was determined how many times later authors and articles have cited a paper of interest. Given the IPL nature of the number of citations, as we pointed out in earlier chapters, an apparently small number of citations is sufficient to discriminate between research articles that remain obscure versus those that become highly visible and eventually initiate and/or direct a subfield of STEM research.

Immediate Visibility (IV): The first measure of visibility employed is the number of citations accumulated in the two years immediately following the year of an article's publication. The number of citations is marked at the publication year plus two years. The shift by two years is intended to mitigate the variations in the timing of an article's publication within the given year. Rather than using this citation count as an indicator, the investigator's perform calculations on the number of citation to explore the article's visibility [100]. They identified the number of citation PDF by article for each PI. This identifies how many of that PI's articles have an unusually large number of citations, indicating an article with a high degree of visibility [61].

Longer-term Visibility (LtV): This indicator is intended to measure lasting STEM interest of an article, as distinct from a flash-in-the-pan transient popularity. To assess the staying power of a paper's visibility the number of citations marked at the publication year plus five years were counted. Again the shift in years is intended to mitigate the variations in the timing of an article's publication. Findings with more enduring significance will tend to garner citations on an ongoing basis year after year, such that those articles may have relatively fewer citations at the two-year mark but a significantly larger number of citations at the five-year mark.

Field Citation Ratio (FCR): It is well known that different STEM disciplines have varying technical standards and traditions concerning the frequency and manner in which articles are cited. Medical journals are notorious for their relatively high citation rates from long citation lists compared with say mathematical journals whose citation rates are much more reserved, given their relatively modest citation lists. To adjust for this social variability the citation number was normalized [100]:

> ...the *Dimensions* database captures all articles related to each field of research each year using the field of research taxonomy from the Australia–New Zealand Science Research Classification scheme, and determines the average number of citations for articles in that field. The citation rate for a given article in that field is compared to the field's average citation rate, and the resulting ratio provides a discipline-neutral citation score.

Influence Indicator

Degree Centrality (DC): The PI's degree of centrality is a measure of the size of the collaboration network of a given researcher. Given a set of articles, a single PI's articles were examined and the number of unique individuals counted who have co-authored one or more papers with that PI. This provides what is called a PI's DC—the size and scope of that STEM researcher's network of collaborators. The DC was calculated for each PI-award observation by looking at the papers associated with an award, and for each funded PI to determine how many unique co-authors appeared across all such papers. This does not indicate the PI's network or influence in relation to the entire community of peer STEM researchers, but does indicate a PI's success in attracting collaborators for work related to the PI's ARO award.

As with visibility, the PI's degree of centrality is an indirect rather than a direct measure of influence. It does require some interpretation based on its context. In some STEM research subdisciplines articles traditionally have hundreds of co-authors, which left unadjusted would inflate degree centrality scores for co-authorship. By limiting the research fields represented in datasets analyzed such discipline-specific influences on this indicator have been mitigated.

Persistence Indicator

Journal Influence (JI): In addition to raw citation analysis, instances of ARO-funded publications that appear in high-impact journals are sought. Because Journal Impact Factor scores are somewhat unreliable and are recalculated on an annual basis, it was determined that for the purposes of the study to not be an appropriate indicator of influence. Instead, each article associated with a PI-award combination was examined, and identified whether any of those articles were indexed by the family of publications in the *Nature* journal series. As *Nature* attracts high-quality papers and its journals have a relatively low acceptance rate for manuscripts, this indicator was considered a suitable, if approximate, proxy for whether an article met the criteria to be published in a high-profile journal.

Continuity in Effort (CiE): A PI's effort under a single ARO award may generate promising findings, but only to the extent that the findings show that the effort merits additional investigation to validate early findings, or to generate more substantive and definitive results. To find instances where multiple awards to a PI represented elements of an extended research agenda, the similarity in the topics of multiple awards to the same PI using topic analysis was examined. A method to determine research continuity by measuring the similarity in the terms and concepts associated with two or more research awards was used and since the technique is quite elaborate we do not describe it here but refer the reader to the original work for details [100].

6.2. Mission Relevance as a Source of Value

ARO is a mission-oriented organization, and it is expected to direct its funding toward STEM research that advances the missions of the Army, AFC and DEVCOM. Some highly-significant STEM research results may not be considered valuable if they are not relevant to that mission. For example, if the ARO funds a ground-breaking study that has no relationship to, or impact on, future Army capabilities, it will have no effective mission relevance (*MR)* value. In this sense, *MR* value depends in part on whether there is a credible means to connect research results to future operational capabilities. However, the potential

MR value of a breakthrough piece of research may not be immediately evident.

Mission Relevance (*MR*) Indicators: *MR* is the second dimension of interest and is chosen because unlike a general fundamental science agency such as the NSF, which funds research to advance all basic science, ARO funds STEM research with a reasonable expectation that the results will contribute to future expansion or enhancement of the Army's capability to carry out its spectrum of missions. While the ARO also seeks to support ground-breaking STEM research, it does so in a manner consistent with its directive to ensure the Army's future capabilities. Therefore, the research value framework incorporates indicators that a set of research results are likely to find if not near-term then at least long-term application to defense needs.

The members of the Army's R&D system are assumed to be the best qualified to understand what STEM advances have the best chance to expand or enhance Army capabilities. Therefore, the half dozen indicators presented in this section provide measures of evidence that other R&D organizations within the Army, and within the broader DoD, show an interest in research that builds on the advances funded by ARO.

Co-Investment Indicators

Concurrent DoD Investment (CI): In examining the articles published with ARO support, instances where that same PI or a co-author of the same article acknowledged support from another DoD sponsor, such as the ONR, AFOSR or DARPA, were identified. Such concurrent investment served as evidence that other DoD research sponsors found the effort relevant to defense missions. For this indicator, the number of unique DoD research sponsors who funded articles produced by the PI that acknowledged the PI's ARO award was determined.

Follow-on DoD Investment (FI): By extracting acknowledgements from articles that a PI co-authored after the award funding ended, it was determined if defense-related organizations supported later research by the same PI. Databases of DoD research funding awards, such as the database maintained by the *Defense Technical Information Center* (DTIC), were searched to see if the PI received such later funding awards. The acknowledgements from articles were then used to extract a list of sponsors for articles published after the end of an award's funding period, and determined how many DoD research sponsors supported those later publications.

Broader Interest Indicators

Social Media Mentions (SMM): The Altmetric's dataset supplied by DSRSI (an affiliate of the firm *Altmetrics*) provided a count of the instances where publications from ARO-funded research in the supplied datasets appeared in social media postings on Twitter. The PI-award level measure counted the number of article tweets, across all articles related to the PI-award combination [100]. Several studies examined the nature of social media mentions of scientific research, with mixed conclusions, although many of these studies sought to connect Altmetrics to measures of *SS*. One common finding was that research results were mentioned frequently in tweets by academics, but not by non-academic accounts.

Transition Potential Indicator

ARL Co-authorship (ARLC): If a STEM researcher at ARL or another defense laboratory took a strong interest in an academic researcher's work, then the ARL STEM investigator may collaborate with the academic researcher on further research and development. Consequently, the study researchers expected to see instances where a PI in a dataset co-authored a scientific paper with a researcher at ARL or a similar DoD research organization. The PI-award level measure identified whether any articles related to the PI's ARO awards were co-authored with ARL or similar researchers.

Army Citations (AC): If STEM researchers affiliated with the Army published a scientific paper that cited previous ARO-funded work, which implies that those researchers found the earlier work relevant to current work answering Army needs. For each article in the study, DSRSI identified the number of citations the article received by papers authored by Army-affiliated researchers. The PI-award level measure counted the total number of Army citations across all articles associated with the PI-award combination. Note that this includes organizations beyond ARL, as in some cases, staff of Army engineering organizations may have conducted research and cited ARO-supported work.

DoD Citations (DoDC): Similar to the above indicator, DSRSI identified how many times a publication associated with an ARO award in the study dataset received citations from articles co-authored by DoD personnel outside of the Army. This indicator showed that members of the broader DoD STEM community read a PI's research paper, and showed enough interest to cite it in their reference list. The PI-award level measure counted the total number of DoD citations across all articles associated with the PI-award combination.

6.3. Potential Breakthrough as Source of Value

The ARO directs its investments in particular at opportunities for "high risk, high payoff research". Among all available opportunities, ARO PMs prioritize those that are likely to have disproportionate positive benefits to future Army capabilities and to the national defense. However, scientific research is inherently uncertain as is its quality, and there are clear limits to STEM researchers' ability to forecast the precise impact that a research project will have on its environment, even if it is unsuccessful. Therefore, ARO PMs prioritize research that has high potential impact—for example, a project where there is a credible argument that positive findings will provide important new insights that can drive the development of novel and disruptive technologies. Collectively, STEM investigators have limited understanding of why some projects that appear to have high potential impact fail to meet expectations [102]. The pursuit of STEM research opportunities with high potential impact requires the acceptance of a high degree of uncertainty around realized impact.

Breakthrough Potential (*BP*) Indicators: *BP* is the third dimension of our STEM research value framework, which is often associated with *SS*, but the two categories have important distinguishing characteristics. In contrast to highly significant research, *BP* research tends to initially go unrecognized and unappreciated by peer researchers. A common characteristic of such research is that it challenges existing STEM assumptions or paradigms, and in doing so, encounters initial resistance but eventually generates completely new insights that were not previously attainable [103]. As the analysis in the present study includes projects that concluded relatively recently, the research team could not predict *BP* with any degree of long-term accuracy. Instead, both intrinsic factors (the nature of the STEM research topic) and extrinsic factors (whether the STEM research results are being applied to technology development) were examined as aspects that signal *BP*.

Invention Indicators

Patient citation (PC): From the data in the *Dimensions* system, instances when a citation to a paper appeared in the text of a patent were identified. The PI-award level measure used was the number of unique patents that cited an article, across all articles associated with the specific PI-award combination.

Government interest in patents (GiP): When an invention is developed, in part with the support of federal funding, any associated patent application is required to include a government interest statement. The *PatentsView* database [104] contains patents issued since 1976 as well as patent applications filed starting in 2001, and has the ability to search the government interest field. PatentsView was used to find patents that cited Army support with grant numbers corresponding to awards in their database, and then constructed the PI-award level measure by counting how many patents cited the specific award to the PI. This signifies that an ARO award contributed directly to an invention, whereas the patent citation indicator denotes an indirect contribution.

Interdisciplinarity Indicators

Field of research dispersion score (Interdisciplinarity): The approximate proxy measure of Interdisciplinarity (I) relied on a part of the process by which the *Dimensions* data system identified the main field of research in which an article is situated. *Dimensions* auto-classifies articles by conducting machine analysis of the entire database of publications to identify the set of terms, concepts, and semantic features that are identified most uniquely with a given field of research (FoR).

In a number of cases an article contains terms and concepts associated with multiple FoRs. Such associations can be interpreted as a signal that the article spans disciplinary boundaries by integrating concepts from more than one FoR. The PI-award level measure counts how many unique FoRs are assigned across all articles related to the PI-award combination.

Commercial Interest Indicators

Corporate authorship (CA): Corporate researchers frequently collaborate with academic researchers in areas where fundamental STEM research has important commercial implications. The degree to which ARO-funded research intersects with commercial R&D efforts is indicated by the count of the number of articles related to the PI-award combination that were co-authored with a researcher having a corporate affiliation.

Corporate investment (CI): Although instances of corporate co-authorship may reflect research where academic and corporate investigators happen to publish results from unfunded collaborative research, evidence of direct corporate financial support of research shows that the private firm identifies the research effort as an investment with an expected payoff. The PI-award level measure counts the number of unique corporate sponsors acknowledged across all articles related to the PI-award combination.

Corporate citations (CC): When a corporate author publishes a paper that cites academic research, it provides a notable indication of commercial interest. These citations are evidence that in pursuing its own private research interests, the corporation found the academic research findings to be a useful contribution to its own efforts. The PI-award level measure counted the number of times the articles associated with the PI-award combination were cited and where at least one of the co-authors had an industrial tie.

Now let us turn to how these 18 indicators of value were used to address the questions posed by the Pentagon leaders that initiated this study.

6.4. Datasets and Results

The base dataset used in the present study consists of information on research awards from ARO to investigators at U.S. universities and this information associates PIs with the research output produced with ARO financial support. The UARC award selected for study was the *Institute for Soldier Nanotechnologies* (ISN) at MIT. The MURI awards were selected based on their thematic similarity to the SI award program and topics from the ARO *Physics* as well as the *Material Science* directorates. Thus, one might expect to see research emphasizing quantum phenomena, strength of new material composites, as well as other areas of common interest in the SI, MURI and ISN datasets.

A key parameter for this study is the distinction between "highly prominent" and "less prominent" research universities. The university ranking determined by the publication *Times Higher Education* for 2015 was used to divide universities into the two groups, the "highly prominent" (top 20) and "less prominent" (not-top 20) groups. They found that the top 20 universities in this ranking system are fairly stable over time. Few universities either ascend to the top 20, or drop out of the top 20, over any period of less than 10 years.

Components of *SS*: As pointed out earlier, in 1965 de Solla Price [41] observed that citation patterns followed an IPL curve, the Pareto PDF [33]. Consequently, in any STEM FoR, a small number of papers garner a disproportionate share of citations in accordance with what Pareto called the *Law of the Unequal Distribution of Results*. The solid line segment depicted in Figure 6.1 is a fit to a hyperbolic PDF for the case of *Immediate Visibility* (number of citations garnered two years after publication of the paper), see Section 5.3. for the derivation of the hyperbolic PDF. The IPL observed and discussed by Price is an asymptotic approximation to the hyperbolic PDF [11]. The hyperbolic PDF (solid curve) fits the data over a much broader range than does the IPL PDF. However, the power-law index of the IPL PDF was found to be within a few percent of that produced by a fit to the hyperbolic PDF.

Note that the quality of fit found here for the hyperbolic PDF is comparable to the excellent fits obtained in Sections 3.3. and 5.3. for an entirely different dataset for a related measure applied to international academic institutions. This

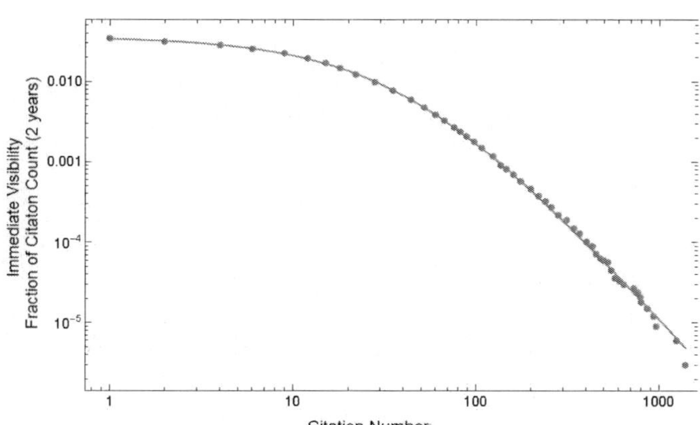

Figure 6.1. The Immedate Visibility component of the *SS* dimension of research value as measured by the fraction of the total number of citations is plotted versus the number of citations. The solid line segment is the fit of a hyperbolic PDF given by Eq.(5.20) to the dataset (dots). The dataset (dots) was generously provided provided by colleagues at RTI.

increases our confidence in the consistency of the ARO dataset with other quality of research value measures obtained independently in the present study.

Just as the Pareto IPL is the asymptotic form of the hyperbolic PDF, the Zipf IPL is the asymptotic form of the MLF CPF and is given by the solid line segment in Figure 6.2. The MLF CPF fits the same dataset used from the analysis in Figure 6.1 over a much broader range of values than does the asymptotic IPL of Zipf. Here again the IPL form of the distribution provides an excellent fit to parameter values using only a fraction of the total dataset.

The comparison between the Pareto IPL and hyperbolic PDFs along with the Zipf IPL and MLF CPFs are given here to emphasize the existence of a theoretical foundation for the measures provided in the present study. The theoretical basis for the IPL distributions used in constructing the measures was not part of the study, but did enter into the interpretation of their results in collaboration with ARO STEM investigators.

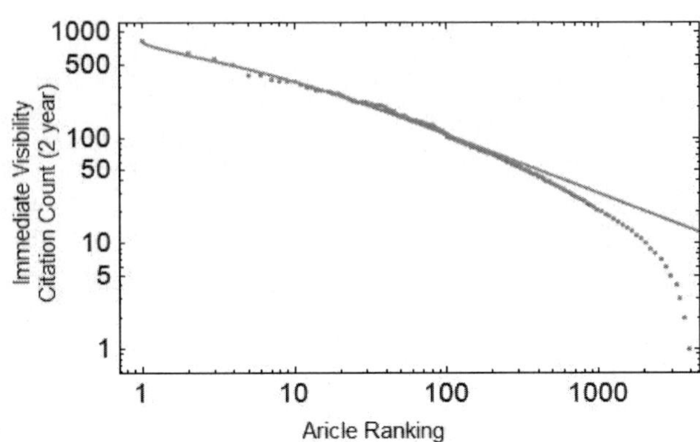

Figure 6.2. The Immedate Visibility component of the *SS* dimension of research value is plotted versus article ranking. The solid line segment is the fit of a Mittag-Leffer CDF to the dataset (dots). The dataset is a reordering of the dataset used in Fig. 6.1.

Relative Measures: The dimensions of research value defined in this chapter are not absolute, but they are systematic. We make use of this systematic behavior to compare the three ARO funding strategies (SI, MURI and UARC) and use the ISN data in each indicator of STEM research value to establish a baseline. The baseline value in each of the 18 indicators is adjusted to zero, which brings each indicator into alignment.

There are six components in each of the three dimensions of STEM research value. Starting at the left column in Table 6.1, see Figure 6.3, the first six indicators are those of *SS* (Scientific significance): IV (Immediate Visibility), number of citations two years after publication; LtV (Longer-term Visibility), number of citations five years after publication; FCR (Field Citation Ratio); DC (Degree of Centrality); JI (Journal Influence) and CiE (Continuity in Effort).

The next six indicators are listed in the column labeled for *MR*: CI (Concurrent DoD Investment); FI (Follow-on DoD investment); SMM (Social Media mentions); ARLC (ARL co-authorship); AC (Army Citations) and DoDC (DoD Citations). The final six indicators are listed in the column for *BP (Breakthrough Potential)*: PC (Patent Citations); GiP (Government Interest in Patents); I (Interdisciplinarity}; CA (Corporate Authorship); CI (Corporate Investment) and

SS	SI-top	SI-nontop	MR	SI-top	SI_nontop	PB	SI-top	SI-nontop
IV	1	1	CI	0	0	PC	-2	-1
LtV	1	1	FI	-2	-2	GiP	-2	-2
FCR	1	1	SMM	0	0	I	-1	0
DC	0	1	ARLC	0	0	CA	0	0
JI	1	2	AC	-1	-1	CI	0	0
CiE	1	2	DoDC	0	0	CC	0	0

Figure 6.3. (Table 6.1): Here we compare the SI funding strategy in both the top and non-top universities to one another in relating how they each compare with the baseline ISN for each of the 18 measures of research value. The ISN baseline value is taken to be 0. There are 6 components in each of the 3 Dimensions of Scientific Value as discussed in the text.

CC (Corporate Citations).

Subsequently, the SI and MURI strategies for both top and non-top universities are computed relative to the ISN baseline value. In this relative view the ISN baseline has the value zero (reference case) given in Table 6.1 with two units above (+1 is somewhat greater and +2 substantially greater than ISN performance) and two units below (-1 is somewhat less and -2 substantially less than ISN performance) giving a range of values of -2 to +2.

SI relative to ISN: In Table 6.1 we compare the 18 indicators of the SI funding at both the top and non-top universities to the ISN baseline. The headings indicate the SIs at top universities and those at non-top universities. All six indicators of *SS* indicate that the SIs do better than the ISN, independently of the ranking of the institution where a STEM investigator conducts the research. In only a single case does the SI coincide with the ISN baseline, indicating equality of STEM research value and that was for SIs at the top universities.

As surprising as the first result is, that the SI strategy results in *SS* research that is overall superior to that of the ISN, it is probably less of a shock than the fact that for three of the six *SS* measures of SI research a non-top universities are superior to those at the top universities. This second counter-intuitive result indicates that SI researchers in non-top universities publish more influential research, based on citations, *h*-indices and other metrics than do such researchers at top universities. Moreover, they follow through with that research for a longer time and are more central players in the resulting research networks than are the ISN researchers in the areas they initiate.

The mission relevance (MR) of the research conducted at both the top and non-top universities relative to the ISN are indistinguishable from one another with the six measured used. The ISN is seen to be clearly superior to the SI-funding strategy in the area of follow-on of DoD funding, as well as in garnering more Army citations.

The breakthrough potential (BP) of the research conducted at both the top and non-top universities relative to the ISN are surprisingly different in the patient citations and interdisciplinary measures where the non-top surpasses the top universities. On the other hand, the SI-funding strategy of all universities does at best no better than ISN in BP research in all six indicators in this dimension of research value. But this may well be a consequence of the ISN being at MIT. In this regard ISN is an anomaly, which is to say that the ISN results are not necessarily representative of potential outcomes at other UARCs, because MIT is itself an anomaly. This observation is a consequence of the fact that MIT has more patents per year than any other U.S. university. It has more patents than the entire University of Texas system but fewer than the entire University of California system, but just barely. The ISN therefore reflects the unique MIT culture in terms of the number of patents received and patents cited and this ratio is perhaps less indicative of a true distinction between the two funding strategies than it is a reflection of MIT's unique status.

Another observation concerns the suppression in MR research having to do with follow-on funding, which reflects the structural difference in the short-term and long-term funding of SI and ISN, respectively. All UARCs have an effectively guaranteed long-term funding, if history is any judge, whereas SI funding has a re-competition every three years. It is a rare SI award that is extended for even two years and this requires compelling experimental or theoretical evidence for its justification. The variance in the follow-on funding has been used to laud one form of funding over another when a less controversial explanation of the variance could well be due to the structural difference having nothing to do with the STEM research quality of the relative outputs.

SS	MURI-top	MURI-nontop	MR	MURI-top	MURI-nontop	BP	MURI-top	MURI-nontop
IV	1	1	CI	0	0	PC	-2	-1
LtV	1	1	FI	-2	-2	GiP	-2	-2
FCR	1	1	SMM	0	0	I	-1	0
DC	0	1	ARLC	0	0	CA	0	0
JI	1	2	AC	-1	-1	CI	0	0
CiE	1	2	DoDC	0	0	CC	0	0

Figure 6.4. (Table 6.2): Here we compare the MURI funding strategy in both the top and non-top universities to one another in relating how they each compare with the baselne ISN for each of the 18 measures of research value. The ISN baseline values and lables are the same as in Table 6.1.

MURIs relative to ISN: In Table 6.2, see Figure 6.4 the 18 indicators of MURI output at both the top and non-top universities are compared to the ISN baseline. Here we see that the MURIs are distinctly superior to the ISN in all six *SS* indicators. However, the evidence is uneven across the *BP* and the *AR* indicators.

We summarize this table as indicating the MURIs at top universities outperform the ISN in *SS*, are equivalent in *BP* and perhaps inferior in *MR* research quality indicators. Even the case where the MURI exceeds the ISN in *MR* it is in the number of mentions on social media which could be accounted for by the superior Public Affairs Offices at the top universities.

Comparing MURIs at top and non-top universities we again encounter something surprising. Here it is clear that the MURIs at the non-top universities in *MR* exceeds the ISN in one indicator, is equal in three, and inferior in two. The MURIs at the non-top universities are superior to the MURIs at the top universities in quality research important to the Army, that is, ARL co-authorship, Army citations and DoD citations, but do not get the credit in social media that the MURIs at the top universities do for work that is less relevant to the Army's immediate needs.

This is where redoing the calculations with the top universities extended from the top twenty to the top twenty-five suppresses the influence of the social media from +2 to 0 (not shown here). Thereby making the MURIs at non-top universities either overall on a par with, or substantially better than, the MURIs at the top universities relative to the ISN. One may conclude from this that the quality of the output of a MURI is virtually independent of the reputation of the host university.

6.5. Summary and Conclusion

I must emphasize that the conclusions expressed in this final section are my own and do not necessarily reflect the inferences drawn by the *RTI International* research team that conducted the study, nor are my conclusions necessarily in conflict with theirs [100]. Given my position as ST and the ARO *de facto* PM of the study I feel obliged to say in the clearest possible terms that the official government position of ARO with regard to the questions addressed herein have not influenced my scientific judgement. I will admit to a certain fondness for ARO as an organization and pride for having served as an ST there for nearly a quarter of a century, but this has not influenced my stated conclusions. On the other hand, it may have cautioned my choice of words in stating those conclusions.

In summary, all evidence from the present study indicates that a long-term geographically restricted funding strategy such as at a UARC could have significantly lowered the research value of the Army's investment in general and to the ARO's investment in particular. The patterns that emerged in the data analysis suggest SI awards contain distinct value different from the MURI and ISN funding mechanisms and are an appropriate primary mechanism for funding new ideas in fundamental STEM research with potential long-term contribution to Army capabilities.

The study was designed to answer two central questions regarding how to best distribute government resources and achieve the 'biggest research bang for the buck', with as little ambiguity as possible. At the risk of being redundant I will spell out what I concluded from the study in answer to the two questions posed.

Question One

The first STEM research question posed in the Pentagon interview discussed in Chapter 1 addressed in this study is:

If ARO had directed its funding primarily to researchers at top universities by redistributing funds now going to non-top universities would ARO have derived higher research value from those investments?

The quantitative measures, while not providing an irrefutable answer to this question, do offer four indications as to what is the right answer:

1) Research funding to top universities – on the basis of institutional reputation alone – is no more likely to generate valuable research results than funding to non-top universities. The data analyses indicate that the value of the research output is essentially independent of the host university. Consequently, the implicit assumption so often made that a research proposal from a top university contains better STEM research ideas than those contained in a proposal from a non-top university is not borne out by the data. The weight of data analyses also rejects the companion assumption that a potential PI from the top university is more capable of bringing his/her research ideas to fruition than a potential PI at a non-top university.

2) The STEM reputation of the university where the PI is employed is less important than other factors in determining the research value of that person's output. This was shown to be the case for both the SI and MURI funding strategies.

3) Research funded through the UARC mechanism generated output whose relative value differs in some respects from STEM research funded through SI and MURI awards. There was however no indication of a systematic increase in research value across the board for the distributed funding mechanism with respect to either of the alternative strategies. This observation is a consequence of the multi-dimensional character of the STEM research value being measured.

4) No funding mechanism shows systematic evidence of producing higher-value STEM research output than any other mechanism across the full suite of 18 indicators.

It is evident that the patterns emerging from the 18 indicators defined, strongly suggest that the implicit assumptions made in order to support the localized (UARC) as opposed to the distributed (SI) funding strategies are not valid. The first question was designed to probe the effect of redirecting the funds to primarily supporting top universities thereby restricting support to all others. Would the overall value of that research to ARO increase?

Each of these four results supports the conclusion that restricting funding to top STEM universities would have slowed the development, if not lost, new STEM research by not funding researchers at a broad spectrum of university rankings. The four conclusions support the inference that the overall STEM value to ARO research would be diminished without the present diversity of funding strategies. Each strategy supports a different set of values from the 18 indicators.

Question Two

The second STEM research question posed in the Pentagon interview discussed in Chapter 1 addressed in this study is:

> If ARO had distributed more research funding through the UARC funding mechanism and less through SI and MURI awards, would it have received higher value from those investments?

The quantitative measures, while not providing a definitive answer, does offer at least three indications as to what is the right answer to the second question:

> 1) PIs at non-top universities produced results of comparable STEM value to those of PIs at top universities, but the value is different in nature. Consequently, the STEM research value changed in character on average when funding a top versus a non-top university, but one cannot say the research value increased or decreased overall using the existing measures.
>
> 2) MURI awards to PIs at non-top universities generated STEM outputs with more immediate application than the MURI awards to PIs at top universities. This counter-intuitive result suggests that if

a PM, during the MURI yearly review process, nudged the attendees in a judiciously chosen direction, such a prod could result in research output being produced along a more Army-oriented path.

3) There is no evidence that the overall STEM research value to ARO is likely to be greater from localized UARC-like awards relative to the distributed SI and MURI awards.

Here again these three results are further support of the conclusion that restricting funding to a small number of top universities with large grants would miss some key research results that could transition faster to application but would not necessarily have gained other results that were higher in other research value indicators.

Main Conclusion:

The patterns that emerge from the data analyses suggest SI awards remain an appropriate primary mechanism for funding new ideas in fundamental STEM research with potential long-term contribution to Army capabilities. It must be borne in mind that this conclusion is reached using a limited dataset and relative, not absolute, measures. A purist will say that we have not given definitive answers to the two questions, and with that we agree. However, all the evidence is consistent with the conclusions reached and are consistent with a negative response to both questions posed in the Pentagon article.

If ARO had directed its funding primarily to researchers at top universities by redistributing funds now going to non-top universities it would have lost STEM research value from those investments.

If ARO had distributed more research funding through the UARC funding mechanism and less through SI and MURI awards, it would have received lower STEM research value from those investments.

Chapter 7

Closing Observations

One of the key questions addressed herein is how to interpret the statistics of nonsimple research networks that are scale free, and formed from citations, collaborations, co-citations, the credentials of those citing, and so on. We acknowledged up-front that the statistical processes with heavy-tailed distributions, those described by IPL PDFs, behave very differently from those described by Gaussian (Normal) PDFs. Most researchers are only vaguely aware that the number of citations to a STEM article, the number of papers published by STEM researchers, as well as the time interval between publications are all IPL PDFs. But even of those that know the statistics are IPL and not Gaussian most have not concerned themselves with what is entailed by those differences in PDFs.

The nonsimplicity of a network is what gives rise to an IPL PDF and the IPL index provides a measure of the level of that nonsimplicity. An IPL in time between events, say an event is the publication of a STEM paper, indicates that the time intervals between the publication of STEM articles are intermittent, which is to say they are fractal in time. On the other hand, suppose the IPL is the number of events with the events being citations to an individual paper. In this case, if the average number of cites received is the same as the average number of citations obtained by the STEM community as a whole, the individual would be a member of the 4-percentile group, as discussed in Chapter 2. This would mean that such an individual is cited more times than 96% of STEM researchers and consequently to be average is to be truly exceptional when the PDF is IPL.

Thus, one cannot interpret the average as an accurate measure of a nonsimple process and we have sketched herein several situations in which doing so has led to catastrophic failures.

Barabási in his book *Bursts; The hidden pattern behind everything we do* [105] interweaves our present-day understanding of the science of human behavior with the story of a burst of human activity – a medieval crusade in his homeland, Transylvania. His ability to relate apparently independent nonsimple phenomena is a consequence of his recognizing the ubiquity of IPL PDFs reflecting the universality of human behavior. He emphasizes that the very essence of IPL PDFs is that they naturally predict rare events and consequently when they are present we can be assured there will be bursts [105]:

> In the end, the patterns of _ follow an inner harmony, short and long delays mixing into a precise law, a law that you probably never suspected you were subject to, that you never made an effort to obey, and that you most likely never knew existed in the first place.

The blank space in the quote replaces the words our e-mailings and the quote interprets what is entailed by the fact that the empirical PDF in the time intervals between e-mailings of an individual is an IPL. If we inserted the words the behavior of STEM investigators into the blank space the interpretation would be equally valid for the communication of STEM research in journals, and in the formation of nonsimple STEM networks. In doing this we have brought the activity of STEM research into the domain covered by Barabási's book. The bursty pattern of human behavior is neither totally random nor completely deterministic, but includes a balance between the two, as determined by the IPL in virtually every measure of STEM research.

The chapter on *The Study* applies the insight gained in the other chapters to the potential measures considered by the RTI research team to quantify the quality of the ARO-supported STEM research. In summary, all evidence from the RTI study indicates that restricting the funding strategy to top STEM research universities would significantly lower the STEM research value of the investment to ARO in particular and to the Army in general. In other words, if the budget for the distributed SI funding were cut in half and invested in the localized funding such as in a UARC this would result in a disproportionate loss in STEM research value to ARO. The patterns that emerge using the quality measures suggest SI awards contain distinct research value different from other funding mechanisms and are an appropriate primary mechanism for funding new ideas in fundamental STEM research with potential long-term contribution to Army capabilities.

The quality of STEM research can only be determine by objective measures. These measures are determined by the number of STEM proposals, their critiques by academic reviewers, the subsequent number of those projects receiving grants, the papers published under these grants, the number of citations each such paper receives not just over the lifetime of the project but long after the project is completed, the number of patients applied for, inventions emerging based on the research, etc.

Herein three dimensions of STEM research quality were identified, each dimension consisting of six distinct measures. These eighteen measures taken together are intended to answer the question as to whether changing the distribution of resources from distributing them widely as determined by the evaluation of individually developed STEM research ideas is superior to block-funding to a top level university in the form of a UARC.

The patterns emerging from the 18 indicators defined, indicate that the implicit assumptions made in order to support the localized as opposed to the distributed funding strategies are not valid. The first question was designed to probe the effect of redirecting all the funds to top universities to the exclusion of all others. Would the overall value of that research to ARO increase?

The results of analysis indicate that the value of the proposed research topic and the likelihood of its eventual resolution is, by and large, independent of the institution housing the STEM researcher proposing the idea. Consequently, the assumption that the ability of a researcher is tied to the reputation of the institution to which the reseacher belongs is a misreading of cause and effect. In point of fact, some of the best research ideas have originated with STEM researchers working in non-top universities as found in the study.

Another interesting observation is that the best research from an Army perspective often come from researchers at non-top universities. One interpretation of the reason for this apparently counterintuitive observation is that talented STEM individuals at non-top universities are typically more open to the suggestions of their PMs as well as to the critiques of other STEM investigators at DoD laboratories. The latter investigators have ‘hands-on’ experience with the needs of the military and often see connections between an Army need and a line of research not visible to the uninitiated.

Each of the results obtained from the study supports the conclusion that restricting funding to top universities would have slowed the development, if not lost, new research by not funding STEM researchers at a broad spectrum of university levels. The results support the conclusion that the overall STEM research value to ARO research would be diminished without the present diversity of funding strategies. Each funding strategy buttresses a different configuration of values from the 18 indicators.

Personal Observations

The conclusions summarized in this chapter is consistent with my personal experience of the vast majority the STEM researchers I have worked with from my first job at a small research company, Physical Dynamics Inc. in 1972, to my last, retiring from the Army Research Office as the ST in Mathematics in 2021. My 50 years of research experience, 18 years at multiple research positions in the private sector, 10 years as a university Physics Professor, and 22 years as a Senior Scientist in Mathematics at ARO, have taught me one thing about research. STEM researchers early on in their careers decide if it is a job or a calling. Consequently, I close my remarks by describing the categories I have consciously and unconsciously, but most importantly without their knowledge, put my colleagues into over the past half century and leave it to the reader to draw the connection between these categories and the findings of the study.

The order of presentation of the personality types has a logic of its own as will become apparent through their descriptions. In any event I will begin with that which appears to be the farthest from STEM research and I have given the label of Sleepers, followed by Keepers, leading into Leapers, who are supported by Creepers, and finally those whose names appear in the popular press, the Reapers. These types are intended to be descriptive, not pejorative and are included here to emphasize that one can seriously discuss an important topic without abandoning one's sense of humor.

STEM research types The five types of research scientists listed here are my idiosyncratic categories for the different ways STEM investigators carry out the various functions associated with STEM research [1]. A perfectly balanced individual would have each characteristic in equal measure, which would form a pentagon with five equal sides. But this balanced individual would not necessarily be the perfect STEM researcher. I will leave it to the reader to decide what constitutes the optimal composition of a researcher, or whether such an optimum actually exists, after they read the definitions.

Sleepers are individuals that have been trained in one or more of the STEM disciplines, but for a variety of reasons have abandoned actively contributing to new STEM knowledge, which is to say they no longer apply the scientific method to experiment or theory to generate new knowledge. They often find satisfaction in teaching, the passing along to the next generation what they and others have learned. Although they themselves no long participate in research that produces new knowledge, as teachers they organize, categorize and put into a logical framework the rather disordered research contributions that others continue to make. Some who have not been gifted with the teaching gene elect to pursue the bureaucratic route of science administration along the academic line of becoming Department Chair, then Dean and so on. The right person in a sensitive position at the proper moment in history can make a tremendous difference in the future direction of STEM research. One such person was V. Bush, the founding Director of OSRD at the entry of the United States into the Second World War, and who, having deep insight into the workings of STEM research, through his report [2] and the BK-debates is considered by many to be the ideological architect of postwar STEM policy.

Of all the STEM areas from which an investigator may choose to continue working once they have decided to stop doing research, in my opinion, teaching is generally the most important.

Keepers are those researchers that refine, redo and carry out experiments that add to the next significant figure in the fine structure or gravitational constants. On the theoretical side these investigators establish new proofs for physical models of well-understood phenomena. As a group these STEM investigators put forward all the objections to new theories and explain in extraordinary detail why this particular experiment is wrong or that proof is flawed. These

---

[1] The five categories of STEM investigators are drawn from my contribution to a Festschrift I edited for my friend and colleague Paolo Grigolini on the occasion of his 80th birthday. They have been sutably modified for the present discussion.

are the maintainers of consistency, searching for the internal contradictions in theories and experimental results; these are the keepers of the status quo.

A Keeper's demand for consistency has a downside which goes by many names. A particularly recent name is a *contemporary pathology of science* [106], which suggests that "previous knowledge" drastically limits innovative thinking in any of the STEM disciplines. These authors believe that true innovation suffers from a too tight embrace with a too big and too flexible corpus of previous knowledge. In some respects a Keeper appears to be the least agreeable of the psychological types, because this is the person who points out your misuse of a term and seems to enjoy finding errors in your work, real or imagined.

Leapers are the type of STEM researcher defined in myth. They are individuals that against all odds and popular STEM belief create the new laws, have the penetrating insights that change a discipline forever, who function completely outside the circumscribed rules of traditional STEM disciplines.

The mathematical physicist Mark Kac, once remarked in talking about such scientists as Sir Isaac Newton and the Noble Laureate Richard Feynman, that they were magicians. He explained that we could do what a genius did if only we were a lot smarter. However, Kac maintained that it was not clear how we could ever do what a magician does. These scientists make connections within and among phenomena in ways that logic cannot readily follow and that often take Creepers many years to reliably test. They are the adventurous spirits who leap to conclusions that solve difficult STEM problems, or leap across STEM disciplines taking an insight from one discipline to make a remarkable but unfounded contribution in another discipline.

Creepers are the worker bees of STEM research. They are the investigators that take us from what is presently known, to what has been hypothesized, conjectured, and/or predicted, but in any event is not yet established. Creepers carry out the systematic investigations that build out and away from our present understanding to what has been guessed, divined or postulated through intuition and/or extrapolation of theory. These are the experimenters and theoreticians that fill in the blank spaces on the landscape of the STEM frontiers, the mapmakers that show us how to creep (cautiously make our way) from one oasis of knowledge established by a Leaper to the next and in doing so make permanent the contributions made for the benefit of all the other STEM investigators.

Creepers are brilliant people. They do not have the magic of the Leapers, but they build the infrastructure that enables the rest of us to understand the STEM breakthroughs revealed through the magic. I mentioned Mark Kac and his remark about some scientists being magicians and others only being very clever. Mark retired from Rockefeller University in the middle 1980s and took a position at the University of Southern California. Once a week he would fly down to San Diego and visit the La Jolla Institute, where I was Associate Director. Since I lived nearest the airport it fell to me to pick him up and take him to our offices in La Jolla, which I did with pleasure.

Mark was a short, round man, with a white fringe around his shining cranium. He had a perpetual smile on his face and was always on the verge of laughing, particularly about some problem in science. He remarked more than once that his entire career had been focused on understanding what we mean by statistical independence. One of his remarkable papers, still frequently quoted after three-quarters of a century is: "Can you hear the shape of a drum?" He was a physicist/mathematician who pursued new problems the way some people go after a good meal, with appetite and good fellowship. The intellectual connections he made in statistical physics and quantum mechanics are continually being rediscovered by each new generation of physicists. He avoided the fashionable and modern, preferring to stay with the fundamental. He thought about such things as the central limit theorem, how it is violated and what the implications of such violations might be. His eyes would dance when he talked about the distribution of eigenvalues in quantum mechanical systems, but no more than when he talked about music or art, or a good book he had just read.

Reapers are those STEM researchers that seem to acquire, sometimes earned and sometimes not, all the accolades and recognition afforded a specific STEM discovery or theoretical insight. It is very often not the originator of a STEM theory, nor the first to perform an extraordinary experiment, who is recognized for their original contribution. It is often the last person to coherently discuss a new phenomenon, not the first, who is credited with its understanding. This does not imply any deception on the part of the Reaper, although that does occasionally happen, but merely points out one of the foibles of being human. Apparently, we prefer to give the credit for an idea, discovery or STEM breakthrough to a single individual rather than to a group or to a sequence of individuals, and the most recent is usually the most obvious recipient of our attention.

Einstein functioned as a Creeper in his five papers on the nature of physical diffusion, except for the first where he was a Leaper. While it is true that his 1905 work on diffusion is completely original, this work was not the first to accurately describe diffusive phenomena. A French physicist named Bachelier, a student of Poincaré, published a paper in 1900 on the diffusion of profit in the French stock market. Bachelier developed the same equations and found the same solutions to those equations, as did Einstein five years later, but unfortunately he was working in a research area not sufficiently mature to recognize what he had done. Consequently Einstein was also a Reaper with regard to recognition that, one could argue, should have more properly have gone to another. It is not that Einstein sought such recognition, it was thrust on him and as a consequence it was denied to another. But in an even greater sense naming diffusion Brownian motion ascribes to Robert Brown much more than he would have willingly accepted, particularly since he was not the first to describe this behavior. It was the Dutch physician, Jan Ingenhousz (1785), who first observed that finely powdered charcoal floating on an alcohol surface executed a highly erratic random motion.

These five categories constitute what I have observed to be the STEM personality types that researchers adopt in establishing and developing collaborative research networks. I leave it to some bright student working on what might be entailed by the interaction of such types with a view to optimizing the value of STEM research

The most successful networks may well be those that follow a dictum that according to the *Reaganfoundation.org*, was on a plaque President Ronald Reagan kept on his desk in the Oval Office of the White House:

There is no limit to what a man can do, or where he can go, if he doesn't mind who gets credit.

References

[1] T. Ferris, *The Science of Liberty; Democracy, Reason, and The Laws of Nature*, Harper/Harper Collins, N.Y. (2010)

[2] V. Bush, *Science, The Endless Frontier*, http://www.nsf.gov/ od/lpa/nsf50 /vbush1945.ht.

[3] D. Sarewitz, "Donald Trump's appeal should be a call to arms", *Nature* **536**, 7 (2016).

[4] D. Wang and A.-L.Barabási, *The Science of Science*, Cambridge University Press, Caambridge, UK (2021).

[5] J.H. Lienhard, "Some ideas About Growth and Quality in Technology", *Technological Forecasting and Social Change* **27**, 265-281 (1985).

[6] J.H. Lienhard, "The Rate of Technological Improvement Before and After the 1830's", *Technology and Culture* **20**, 515-530 (1979).

[7] D.S.L. Cardwell, *Turning Points in Western Technology*, Neale Watson Academic Publications, New York (1972).

[8] W.B. Arthur, *The Nature of Technology, What it is and how it evolves,* Free Press, NY (2009).

[9] A. Kott, "Toward Universal Laws of Technology Evolution: Modeling Multi-Century Advances in Mobile Direct-Fire Systems", *The Journal of Defense Modeling and Simulation* (2019).

[10] B.J. West, D. West and A. Kott, "Size and history combine in allometry relation of technology systems" , *J. Defense Modeling & Simulation,* 1-6 (2020). doi.org/10.1177/1548512920942327.

[11] B.J. West and P. Grigolini, *Crucial Events; Why are catastrophes never expected?,* Studies of Nonlinear Phenomena in Life Science - Vol. 17, World Scientific, NJ (2021).

[12] NAS Assessment of DoD Basic Research (2005). http://www.nap.edu /openbook/0309094437/html/1.html.copyright 2005, 2001.

[13] An Army Material Command Human Resources Representative indicated this phenomenon at an Army-wide Senior Research Scientist (ST) meeting in Warren, MI. January 25, 2005.

[14] US Air Force, Office of the Chief Scientist of the Air Force, "Science and Technology workforce for the 21st Century", Washington D.C., July 1999; US Navy, Office of the Assistant Secretary of the Navy (Research, Development, and Acquisition), Naval Research Advisory Committee, "Science and Technology (S&T) Community in Crisis", Washington, D.C., May 2002.

[15] R. Adrian, *The Analyst: The Mathematical Museum* **1**, 93-109 (1809).

[16] P. Pande and L. Holpp, *What is Six Sigma?*, McGraw Hill, New York (2002).

[17] H.M. Gupta, J.R. Campanha and F.R. Chavorette, *Power-law distribution in high school education: effect of economical, teaching and study conditions*, arXir.0301523v1 (2003).

[18] T. Rose, *The End of Average*, Harper Collins, NY (2016).

[19] B.J. West, C. Arney and K. Hutchinson, *Dialogues Conncening Science, Technology, and Intellect in American Sciety's and Military's Future,* Cambridge Scholars Publishing, UK (2021).

[20] *USAF Aircraft Accidents, February 1950", Accident-Report.com,* http://www.accident-report.com/Yearly/1950/5002.html.

[21] G.S. Daniels, *The "Average Man"?*, TN-WCRD-53-7, Dayton:Wright-Patterson AFB, Air Force Aerospace Medical Research Lab (1952).

[22] A.J. Lotka, "The frequency distribution of scientific productivity", *J. Wash. Acad. Sci.* **16**, 317 (1926).

[23] B.J. West and W. Deering, *The Lure of Modern Science, Fractal Thinking*, World Scientific, NJ (1995).

[24] B.J. West, L.A. Griffin, H.J. Frederick and R.E. Moon, "The indpendently fractal nature of respiration and heart rate during exercise under normobaric and hyperbaric conditions", *Respiratory Physiology & Neurobiology* **145**, 219 (2005).

[25] N. Scaffeta, S. Picozzi and B.J. West, "A trade-investment model for distribution of wealth", *Physica D: Nonlinear Phenomena* **193**, 338-352 (2004).

[26] A.-L. Barabási, *Linked*, Plume, New York (2003).

[27] S. Strogatz, *SYNC*, Hyperion Books, New York (2003).

[28] B.J West, M. Turalska and P. Grigolini, *Network of Echoes: Immitation, Innovation and Invisibe Leaders,* Springer, Switzerland (2014).

[29] A.F.J. van Raan, "Measuring Science", in H.F. Moed et al., Eds., *Handbook of Quantitative Science and Technolgy Research,* pp.19-50, Kluwer Academic Publishers, Printed in the Netherlands (2004).

[30] A.-L. Barabási and R. Albert, "Emergence of scaling in random networks". *Science* **286**, 509–512 (1999).

[31] D.J. Watts, *Small Worlds*, Princeton University Press, Princeton, New Jersey (1999).

[32] B.J. West, *Physiology, Promiscuity and Prophecy at the Millennium: A Tale of Tails*, Studies of Nonlinear Phenomena in Life Science-Vol.7, World Scientific, Singapore (1999).

[33] V. Pareto, *Cours d'Economie Politique* [*Political Economy Course*], Lausanne (1897).

[34] B.J. West, *Simplifying Complexity: Life is Uncertain, Unfair and Unequal,* Benthem Books, Sharja, UAE (2016).

[35] W. Shockley, "On the Statistics of Individual Variations of Productivity in Research Laboratories", Proceedings of the IRE, 270 (1957).

[36] D. Sornette, *Critical Phenomena in Natural Sciences, 2nd Edition,* Springer-Verlag, Berlin (2004).

[37] J. Juran, *Quality Control Handbook,* first published in 1951, 5th Edition, McGraw-Hill, New York (1999).

[38] J. Juran, *Managerial Breakthroughs*, McGraw-Hill, New York (1964).

[39] R. Koch, *The 80/20 Principle: The Secret of Achieving More with Less*, Nicholas Brealey Publishing (1998).

[40] A.J. Lotka, *Elements of Mathematical Biology*, Dover , New York (1956) first published in 1924.

[41] D.J. de S. Price, "Networks of Scientific Papers", *Science* **149**, 510-515 (1965).

[42] S. Redner, "How popular in your paper? An empirical study of the citation distribution", *Eur. Phys. Jour. B* **4**, 131-134 (1998).

[43] F. Liljeros, C.R. Edling, L.A. Nunes Amaral, H.E.Stanley and Y. Aberg, "The web of human sexual contacts", *Nature* **411**, 907 (2001).

[44] M. Granovetter, "The strength of weak ties", *American Journal of Sociology* **78**, 1360-1380 (1973).

[45] G.U. Yule, "A mathematical theory of evolution based on the conclusions of Dr. J.C. Willis, FRS", *Philos. Trans. R. Soc. Lond. B,* vol. 213, 21 (1925).

[46] D.J. de S. Price, "A general theory of bibliometric and other cumulative advantage processes", *J. Amer. Soc. Inform. Sci.* **27**, 292–306 (1976).

[47] D.J. de S. Price, *Little Science, Big Science ... and Beyond*, Columbia University Press, New York (1986).

[48] W. Glanzel, "On the h-index – A mathematical approach to a new measure of publication activity and citation impact", *Scientometrics* **67** (2), 315-321 (2006).

[49] S. Fortunato, C.T. Bergstrom, K. Borner, et al., "Science of Science", *Science* **359**, 1-7 (2018).

[50] A.Zeng, Z. Shen, *et al.*,"The science of science: From the perspective of complex systems", *Phys. Reports* **714**, 1-73 (2017).

[51] V. Bush, "As we may think", *Atlantic Monthly* **176** (1), 101-108 (1945).

[52] K. J. Boudreau, E. C. Guinan, K. R. Lakhani, C. Riedl, "Looking across and looking beyond the knowledge frontier: Intellectual distance, novelty, and resource allocation in science", *Manage. Sci.* **62**, 2765–2783 (2016).

[53] E. Leahey and J. Moody, 'Sociological innovation through subfield integration", *Soc. Currents* **1**, 228–256 (2014).

[54] A. Yegros-Yegros, I. Rafols, and P. D'Este, "Does interdisciplinary research lead to higher citation impact?The different effect of proximal and distal interdisciplinarity", *PLOS ONE* **10**, e0135095 (2015).

[55] T. Jia, D. Wang and B.K. Szymanski, "Quanifying patterns of research-interest evolution", *Nat. Hum. Behav.* **1**, 0078 (2017).

[56] M.E.J. Newman, "Coauthorship networks and patterns of scientific collaboration", *PNAS* **102**, 5200-5205 (2004).

[57] A.F.J. van Raan, "Fractal dimension of co-citations", *Nature* **347**, 626 (1990).

[58] S. Redner, "On the meaning of the h-index", *J. Stat. Mech.* L03005 (2010).

[59] J.E. Hirsch, "An index to quantify an individual's scientific research output", *PNAS* **102**, 16569-16572 (2005).

[60] M.M. King, etc., "Men set their own cites high: Gender and self-citation across fields and over time", *SOCIUS* **3**, 1-22 (2017).

[61] R. Sinatra, D. Wang, P. Deville et al., "Quantifying the evolution of individual scientific impact", *Science* **354**, aaf5239-1 (2016).

[62] A.F.J. van Raan, "Fatal Attraction: Conceptual and methodological problems in the ranking of univeristies by bibliometric methods", *Scientometrics* **62**, 133-143 (2005).

[63] H.F. Moed, W.J.M. Burger et al., "The use of bibiometric data for the measurement of university research performance", *Research Policy* **14**, 131-149 (1985).

[64] D.A. King, "The scientific impact of nations", *Nature* **430**, 311-316 (2004).

[65] J. Adams, "The rise of research networks", *Nature* **490**, 335-336 (2012).

[66] J. Molinari and A. Molinari, "A new methodology for ranking scientific institutions", *Scientometrics* **75**, 163-174 (2008).

[67] A. Chatterjee, A. Ghosh and B.K. Chakrabarti, "Universality of Citation Distributions for Academic Institutions and Journals", *PLoS ONE* **11**(1): e0146762 (2016).

[68] Z. Néda, L. Varga and T.S. Biró, "Science and Facebook: The same popularity law!", *PLoS ONE* **12**(7): e0179656 (2017).

[69] B.J. West and P. Gigolini, *Complex Webs, Anticipating the Improbable,* Cambridge University Press, Cambridge, UK (2011).

[70] B.J. West, K. Mahmoodi and P. Grigolini, *Empirical Paradox, Complexity Thinking and Generating New Kinds of Knowledge,* Cambridge Scholars Publishing, UK (2019).

[71] J.G. Foster, A. Rzhetsky and J.A. Evans, "Tradition and Innovation in Scientists' Research Strategies", *Am. Soc. Rev.* **80**, 875882 (2015).

[72] B. Uzzi, S. Mukherjee, M. Stringer and B. Jones, "Atypical Combinations and Scientific Impact", *Science* **342**, 468-472 (2013).

[73] Y-B. Zhou, L. lu and M. Li, "Quantifying the influence of scientists and their publications: Distinguish prestige from popularity", *New J. Physics* **14**, 033033 (2012).

[74] C. Bergstrom, "Eigenfactor: measuring the value and prestige of scholarly journals", *C&RL News* **68**(5), 214-316 (2007).

[75] P. Meakin, *Fractals, scaling and growth far from equilibrium*, Cambridge University Press, Cambridge, UK (1998).

[76] N. Goldenfeld, *Lectures on phase transitions and the renormalization group*, Addison-Wesley Pulblishing, Reading , MA (1992).

[77] T. Mora and W. Bialek, "Are Biological systems Poised at Criticality?", *J. Stat. Phys.* **144,** 268-302 (2011).

[78] L.H. Sullivan, "The tall office building artistically considered", *Lippencott Magazine* (March), 403 (1896).

[79] B.J. West, *Nature's Patterns and the Fractional Calculus,* DE Gruyter, Berlin (2017).

[80] A.F.J. van Raan, "Universities Scale Like Cities", *PLoS OnE* **8**(1), e59381 (2013).

[81] L.M.A. Bettencourt, J. Lobo, D. Helbing, C. Kuhnert and G.B. West, "Growth, innovation, scaling and the pace of life in cities", *PNAS* **104**, 7301 (2007).

[82] L.R. Taylor, "Aggregation, variance and mean", *Nature* **189,** 732-735 (1961).

[83] D. West and B.J. West, "Stochastic origin of allometry", *EPL* **94,** 38005 (2011).

[84] W.W. Calder III, *Size, Function and Life History*, Harvard University Press, Cambridge, MA (1984).

[85] K. Schmidt-Nielsen, *Scaling, Why Is Animal Size so Important?*, Cambridge University Press, Cambridge (1984).

[86] J. Brownlee, "Density of death rate: Farr's Law", *J. Roy. Soc. Stat.* **83**, 280-283 (1920).

[87] J.T. Hack, "Studies in longitudinal profiles in Virgina and Maryland", *US Geol. Surv. Peof. Pap. B* **294,** 52 (1957).

[88] B.B. Mandelbrot, *Fractals, Form and Chance*, W.H. Freeman San Francisco, CA (1977).

[89] G. West, J.H. Brown and B.J. Enquist, "A general model for ontogenetic growth", *Nature* **413**, 628 (2001).

[90] A.J. Kerkhoff and B.J. Enquist, "The implications of sclaing approaches for understanding resilience and reorganization in ecosystems", *BioScience* **57**, 489-499 (2007).

[91] C. Cempel, M. Tabaszewski, and S. Ordysinski, "Allometry scalling and accidents at work", *Int. J. Occup. Safety and Ergonomics* **22,** 173-178 (2016).

[92] A. Kott, S. Gart and J. Pusey, "From cockroaches to tanks: the same power-mass-speed relation describes both biological and artificial ground-mobile systems", *PLoS ONE* **16**(4): e0249066. https://doi.org/10.1371/journal.pone.0249066.

[93] B.B. Mandelbrot, *The Fractal Geometry of Nature*, W.H. Freeman, San Francisco, CA (1982).

[94] A.L. Kinney, "National scientific facilities and their science impact on non-biological research", *PNAS* **104**, 17943-17947 (2007).

[95] A. Molinari and J. Molinari, "Mathematical aspects of a new criterion for ranking scientific institutions based on the h-index", *Scientometrics* **75**, 339-356 (2008).

[96] M. Turalska and B.J. West, “Fractional Dynamics of Individuals in Complex Networks”, *Front. Phys.* 6:110, (2018). *doi: 10.3389/fphy.2018.00110.*

[97] B.J. West, *Fractional Calculus View of Complexity: Tomorrow's Science*, CRC Press, boca Raton, FL (2016).

[98] F. Mainardi, “Why the Mittag-Leffler Functon Can Be Considered the Queen Function of the Fractional Calculus?”, *Entropy* **2020**, 22, 1359-1378 (2020). *https://doi.org/10.3390/e22121359.*

[99] L.A. Adamic and B.A. Huberman,“Power-law distribution of the World Wide Web”, *Science* **287**, 2115 (2000).

[100] J. Alexander, *Army Research Office Funding Strategy and Research /assessment: Methodology Report*, RTI International, Research Triangle Park, NC (2021).

[101] D. Hicks, P. Wouters, L. Waltman, S. de Rijcke & I. Rafols, “The Leiden Manifesto for research metrics”, *Nature* **520**(7548), 429–431 (2015).

[102] C. Wagner and J.M. Alexander,“Evaluating transformative research programmes: A case study of the NSF Small Grants for Exploratory Research programme”, *Research Evaluation* **22**(3):187-197 (2013).

[103] National Science Board, *Enhancing Support of Transformative Research at the National Science Foundation.* National Science Foundation (2007).

[104] PatentView data base, https://www.patentsview.org/

[105] A.-L. Barabási, *Bursts; The hidden pattern behind everything we do*, Dutton, New York (2010).

[106] C. Modonesi, L. Farina, I. Licata, R. Germano, J.P. Zbilut and A. Giuliani, “A contemporary pathology of science”, *Ann Ist Super Sanita* **44**, 211-213 (2008).

Index

About the Author

Bruce J. West
Visiting Scholar
Center for Nonlinear Science,
University of North Texas, Denton, TX, USA
and
and Office of Research and Innovation,
North Carolina State University, Rayleigh, NC, USA
Email: brucejwest213@gmail.com